Eco-Standards, Product Labelling and
Green Consumerism

## Consumption and Public Life

Series Editors: **Frank Trentmann** and **Richard Wilk**

*Titles include:*

Mark Bevir and Frank Trentmann (*editors*)
GOVERNANCE, CITIZENS AND CONSUMERS
Agency and Resistance in Contemporary Politics

Magnus Boström and Mikael Klintman
ECO-STANDARDS, PRODUCT LABELLING AND
GREEN CONSUMERISM

Daniel Thomas Cook (*editor*)
LIVED EXPERIENCES OF PUBLIC CONSUMPTION
Encounters with Value in Marketplaces on Five Continents

Nick Couldry, Sonia Livingstone and Tim Markham
MEDIA CONSUMPTION AND PUBLIC ENGAGEMENT
Beyond the Presumption of Attention

Amy E. Randall
THE SOVIET DREAM WORLD OF RETAIL TRADE AND
CONSUMPTION IN THE 1930S

Kate Soper and Frank Trentmann (*editors*)
CITIZENSHIP AND CONSUMPTION

Harold Wilhite
CONSUMPTION AND THE TRANSFORMATION OF
EVERYDAY LIFE
A View from South India

*Forthcoming:*

Jacqueline Botterill
CONSUMER CULTURE AND PERSONAL FINANCE
Money Goes to Market

Roberta Sassatelli
FITNESS CULTURE
Gyms and the Commercialisation of Discipline and Fun

---

Consumption and Public Life
Series Standing Order ISBN 978–1–4039–9983–2 Hardback
978–1–4039–9984–9 Paperback
(*outside North America only*)

You can receive future titles in this series as they are published by placing a standing order. Please contact your bookseller or, in case of difficulty, write to us at the address below with your name and address, the title of the series and the ISBN quoted above.

Customer Services Department, Macmillan Distribution Ltd, Houndmills, Basingstoke, Hampshire RG21 6XS, England

---

# Eco-Standards, Product Labelling and Green Consumerism

Magnus Boström
*Södertörn University College, Sweden*

Mikael Klintman
*Lund University, Sweden*

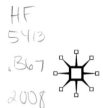
First published 2008 by
PALGRAVE MACMILLAN

Palgrave Macmillan in the UK is an imprint of Macmillan Publishers Limited, registered in England, company number 785998, of Houndmills, Basingstoke, Hampshire RG21 6XS.

Palgrave Macmillan in the US is a division of St Martin's Press LLC, 175 Fifth Avenue, New York, NY 10010.

Palgrave Macmillan is the global academic imprint of the above companies and has companies and representatives throughout the world.

Palgrave® and Macmillan® are registered trademarks in the United States, the United Kingdom, Europe and other countries.

ISBN-13: 978–0–230–53737–8 hardback
ISBN-10: 0–230–53737–5 hardback

This book is printed on paper suitable for recycling and made from fully managed and sustained forest sources. Logging, pulping and manufacturing processes are expected to conform to the environmental regulations of the country of origin.

A catalogue record for this book is available from the British Library.

Library of Congress Cataloging-in-Publication Data
Boström, Magnus.
    Eco-standards, product labelling and green consumerism / Magnus Boström, Mikael Klintman.
        p. cm.—(Consumption and public life)
    Includes bibliographical references.
    ISBN-13: 978–0–230–53737–8
        1. Eco-labeling. 2. Green products. 3. Green marketing. I. Klintman, Mikael, 1968- II. Title.

HF5413.B67 2008
658.8′02—dc22                                    2008020597

10  9  8  7  6  5  4  3  2  1
17  16  15  14  13  12  11  10  09  08

Printed and bound in Great Britain by
CPI Antony Rowe, Chippenham and Eastbourne

# Contents

# List of Figures and Tables

## Figures

## Tables

# Acknowledgements

This book is based on research projects financed by the Bank of Sweden Tercentenary Foundation, the Swedish Research Council, the Swedish Research Council for Environment, Agricultural Sciences and Spatial Planning, and the Knut and Alice Wallenberg Foundation. We thank them all.

We want to acknowledge the contributions made by our colleagues at the Stockholm Centre for Organizational Research, Stockholm University; at Stockholm University College; at the Research Policy Institute, Lund University; and at Massachusetts Institute of Technology in Cambridge, MA. Many people at these institutes and elsewhere have contributed generously with their inspiration and advice. In particular, we wish to thank, in alphabetical order.

Andy Zachs, Anna-Lisa Lindén, Anna Olofsson, Annika Kronsell, Beatrice Bengtsson, Boris Holzer, Claes-Fredrik Helgesson, Ebba Sjögren, Erika Jörgensen, Gert Spaargaren, Göran Ahrne, Göran Sundqvist, Ingmar Elander, Karin Bäckstrand, Karin Fernler, Karin Svedberg Nilsson, Kristina Tamm Hallström, Lars Gulbrandsen, Lena Ekelund, Linda Soneryd, Martin Rein, Mats Benner, Maurie Cohen, Michele Micheletti, Nils Brunsson, Renita Thedvall, Rolf Lidskog, Sander van den Burg, Sheila Jasanoff, Sofia Nilsson, Sophie Dubuisson Quellier, Steven Yearley, Ylva Uggla, and Åsa Vifell.

Special debts of gratitude are owed to the many people in several countries who have given freely of their time and effort during our research interviews.

We also thank Nina Colwill and Alan Crozier for their patient work in correcting our English in various chapters of this book.

Furthermore, we would like to mention events and meetings where people have provided us with particularly valuable insights and reinforcing critiques: The Swedish Sociological Association in 2006; The International Sociological Association in June 2006; The XVI World Congress of Sociology in Durban, South Africa, in July 2006; the book seminars at Sundbyholm and at Stockholm Centre for Organizational Research in Sweden in the Spring of 2007, and in Genarp in August 2007; the Nordic Consumer Research Conference in Helsinki in October 2007; the Seminar Series at the Research Policy Institute, Lund University, between 2005 and 2007, and finally the seminar at the Section for x *Acknowledgements*

Science and Technology Studies, University of Gothenburg, in February 2008.

Finally, we would like to express our warmest gratitude to our families – Lotta, Isak, Ivar, Arvid, and Jenny, Leo, Bruno, Fred – for all their encouragement and for providing us with the most important inspiration.

MAGNUS BOSTRÖM AND MIKAEL KLINTMAN
*Stockholm and Lund, Sweden*, April 2008

# List of Abbreviations

| | |
|---|---|
| ACPA | American Crop Protection Association |
| AF&PA | American Forest and Paper Association |
| AI CSRR | Association of Independent Corporate Sustainability and Responsibility Research |
| CSA | Canadian Standards Association |
| DIY | Do-It-Yourself |
| DJSI | Dow Jones Sustainability Indices |
| EMS | Environmental management systems |
| EMO | Environmental movement organization |
| Eurosif | European Social Investment Forum |
| FLO | Fairtrade Labelling Organization International |
| FoE | Friends of the Earth |
| FSC | Forest Stewardship Council A.C. |
| GEN | Global Ecolabelling Network |
| GM | Genetic modification |
| GRI | Global Reporting Initiative |
| ICES | International Council for the Exploration of the Sea |
| IFOAM | International Federation of Organic Agriculture Movements |
| ILO | International Labour Organization |
| IOAS | International Organic Accreditation |
| ISEAL Alliance | International Social and Environmental Accreditation and Labelling Alliance |
| ISO | International Organization for Standardization |
| IUCN | World Conservation Union |
| KF | Kooperativa Förbundet (The Swedish Cooperative Union) |
| LRF | Lantbrukarnas Riksförbund. (The Federation of Swedish Farmers) |
| MAC | Marine Aquarium Council |
| MSC | Marine Stewardship Council |
| NGO | Non-governmental organization |
| NOP | National Organic Program |
| NOSB | National Organic Standards Board |
| OECD | Organisation for Economic Co-operation and Development |

| | |
|---|---|
| OFPA | Organic Food Protection Act |
| OTA | Organic Trade Association |
| PEFC | Programme for the Endorsement of Forest Certification schemes |
| PRI | Principles for Responsible Investment |
| SAI | Social Accountability International |
| SFB | Sustainable Forestry Board |
| SFI | Sustainable Forestry Initiative |
| SIF | Social Investment Forum |
| SLU | The Swedish University of Agricultural Sciences |
| SMO | Social movement organization |
| SRI | Socially Responsible Investing (the basis for green and ethical mutual funds) |
| SSNC | Swedish Society for Nature Conservation |
| TNC | Transnational Corporation |
| UKSIF | UK Social Investment Forum |
| UN | United Nations |
| UNEP FI | United Nations Environment Programme Finance Initiative |
| USDA | US Department of Agriculture |
| WTO | World Trade Organization |
| WWF | World Wide Fund for Nature |

# 1
# Introduction: Green Consumerism, Green Labelling?

> He knows how easy it is to be bad, how one has only to relax for
> the badness to emerge.
>
> (J.M. Coetzee, in *Youth*, 2003: 132)

## Risk culture and eco-standards

As increasingly conscientious 'citizen-consumers', many of us seek
environmentally friendly alternatives to petrol. But how do we avoid
farm workers producing these alternatives in slave-like conditions? We
seek ways of reducing our carbon footprints. But can anyone estimate
such footprints in an accurate way? We hear that it is ecologically sound
to choose goods produced geographically close to us. But would it not
be more humane to support local production in the poorer, southern
countries? We proudly tell our friends that the animals that we eat have
been raised on ecologically responsible farms. But does not all farm-based
meat production use immense amounts of clean water that would be of
better use for humans and for the production of vegetarian food?

   We live in a risk culture. People are overwhelmed with alarms
about food contamination, over-fishing, clear-felling of forests, loss of
biodiversity, climate change, chemical pollution, the potentially
negative consequences of genetically modified products, and many
other environmental and health-related risks. People feel that they
cannot rely merely on public authorities, who have often failed to
shoulder their responsibilities. Many people continue as usual, aware
of the alarms murmuring irritatingly somewhere in the background,
but with a hopeless feeling that their own actions are meaningless –
'just my changing to a green lifestyle won't change the whole picture,
and I will certainly get cancer anyway'. But many other people have

expressed a willingness to make dramatic changes in their everyday lives in order to decrease their ecological footprint. This goal can be partially accomplished through consumer choice. Journalists, religious leaders, teachers, environmental movement campaigners, other policy-makers, and debaters have helped to explain how we can contribute to social change through consumer choice. They tell us that it is no longer merely votes that matter; politically and ethically motivated consumer choice in the market arena matters as well. Citizens express political concerns through more active consumer choices, through 'political consumption' (Micheletti, 2003), either by boycotting products or by 'buycotting' – by consciously choosing environmentally and/or socially friendly products. Yet green choices do not always represent the most inexpensive option, so the consumer who wants to buycott often pays more.

As a response to increasing and widespread concern, to the lack of public trust in existing policies and regulatory arrangements, and to the expressed or imagined willingness among consumers to engage in political consumption, politicians, state agencies, environmental movement organizations, consumer associations, business actors, churches, labour unions, and individual consumers are increasingly engaged in finding and developing new market-based and consumer-oriented instruments. In this book, we are interested in the many voluntary instruments available – eco-standards[1] such as shopping guides, eco-labels, stewardship certificates, ranking and rating, green mutual funds, environmental management systems, environmental declarations, codes of conduct, reporting standards, and certain trademarks with an eco-friendly profile – that have been introduced into the market to address consumers' environmental concerns. There appears to be general agreement among political and other actors across the ideological spectrum in several countries that such instruments are useful and powerful. Such ideals as economic, social, environmental, and democratic values lie beneath their affirmative attitude.

The eco-standards can also be controversial. Green labels, for instance, symbolically distinguish between green and grey, good and bad, sustainable and unsustainable, safe and risky, and such symbolic differentiation has often proved to be highly controversial: 'What is wrong with this (unlabelled) product?' In this book, we argue that green labelling is inherently a 'political' affair – or, as Beck (1992) would say, 'sub-political' – and that the sub-politics concerns both the production and the consumption side of the labelling.

This book draws upon several years of empirical and theoretical research in analysing the practical *tools* of green consumerism, with a focus on green labels (including eco-labels, stewardship certificates, and green mutual funds). By the term *green labels,* we refer to markers that are presented to consumers or professional purchasers, and are assumed to help to distinguish environmentally beneficial consumer choices from 'conventional' ones (see Chapter 3). A main point of departure for our studies has been the many knowledge gaps and ideological diversities that surround environmental issues in general, and green labelling and green consumerism in particular. With the many social and environmental complexities as background, we are interested in why – and how – such green instruments are produced, introduced, and debated, and what preconditions they offer for the greening and democratization of society. In what ways are the social and environmental complexities translated to a simple, categorical – 'this is the green(est) choice' – label? Do the labels correspond to the concerns of the consumers who use them or potentially use them? In what way could such eco-standards assist consumers and the greening of society?

Rather than focusing on consumers' front stage, on their decision-making process, or on their knowledge and awareness of particular eco-standards,[2] we are investigating processes on their back stage, *behind* the final label that is placed on the product or service. The chief objective of this book is to analyse and discuss green consumerism and the setting of eco-standards, with attention to the conditions, opportunities, and dilemmas of green labelling processes. The focus is on the supply side – on the development of instruments for green (political) consumerism and on the relationship between production and consumption. We argue that this book is original in the sense that it analyses green labelling in relation to green consumerism in general – not merely in relation to a specific sector, as has been common in the literature. Hence, we base our analyses on comparisons of labelling projects in various sectors: forestry, paper products, fishery, organic foods, genetically modified foods, green/ethical funds, and green electricity. Our specific focus is on labelling projects in two countries – Sweden and the United States – with examples from other countries. Sweden is examined as an individual nation and as a member of the European Union. Furthermore, we analyse labelling practices in their discursive, organizational, regulatory, political, and transnational contexts, which we claim to be crucial contexts for assessing the potential of green consumerism.

This introductory chapter provides the reader with a presentation of the problem of green consumerism and information instruments aimed at concerned consumers. We develop four general themes in this chapter to guide our discussion in the book. This is followed by an outline of the book, where we briefly introduce key concepts. Finally, we provide some notes on the method we have used and on normative assumptions, and finish with a brief overview of the chapters that follow.

## Four themes: politics, trust, differentiation, and mismatch

### Does politics distort labelling?

There is a common ambition within the sociology of scientific knowledge and within interdisciplinary science and technology studies, which we share. In many such studies, scholars want to stimulate a better public understanding of science by elucidating how 'scientists are neither Gods nor charlatans; they are merely experts, like every other expert on the political stage' (Collins & Pinch, 1993: 145; see also Irwin & Wynne, 1996; Yearley, 2005). Given all social and environmental complexities, including our view of green labelling as inherently political, we have been struck by the recurrent framings of green labels as 'neutral information', 'based on objective knowledge', 'scientifically valid', and so on. In our view, it is also misconceived to reject labelling because it fails to be 'neutral', 'objective', or 'scientifically valid', or because it is based on 'ideological', 'political', 'social', or 'strategic' rationales.

In a case concerning whether or not there should be a mandatory label on products produced through genetic modification (GM), the polarization between science and ideology is conspicuous. As observed by one GM proponent, Henry I. Miller, who has been a major voice against mandatory GM labelling:

> Although exhaustive tests indicated that the milk is no different or less wholesome than that obtained from untreated cows, activists demanded special regulations, including mandatory labelling of dairy products from BST-treated animals. (Miller, 2007: 281)

In his view, the ideological pole is represented by, among others, 'activists', 'anti-biotechnology' groups, and other non-governmental organizations that intend to 'demonize those products [and] to intimidate their producers, distributors and retailers, and to pounce on any inconsequential mislabelling' (Miller, 2007: 281).

Yes, *all* sides of labelling, we maintain, have inherent political, social, ideological, and strategic dimensions. Yet science does indeed play a vital role in labelling. How can all these dimensions be combined?

Labels are categorical claims. Labellers generally claim that labelled products are better for the environment, for health, for animal welfare, for social justice, and so forth, than competing 'conventional' products. Such claims need to be legitimized by reference to authoritative knowledge claims, which are typically provided by science. We have noticed many references to science in the labelling programmes we have studied. Labellers claim, for example, that they use life cycle analysis in order to establish scientifically valid labelling criteria. They establish organizational forums for the systematic inclusion of scientific expertise. They claim that labelling criteria must be based upon scientific evidence in order to be viewed as credible. The word 'science' appears in the vocabulary of most labelling organizations and proponents, in their marketing activities, and in communications among stakeholders. Moreover, the importance of science, objectivity, and independence of vested interests is implied in a plethora of communication documents on labelling, on all sides of the disputes. In the case of organic food in the United States, for example, it is common to read claims such as:

> Organic food is certainly safer and better than the chemical-doused, genetically contaminated, or irradiated food typically found on grocery store shelves. (Organic Trade Association, 2002)

The word 'science' is not mentioned here; yet the statement implicitly stresses the importance of highly sophisticated science, in which the spokespersons seem to hold a high trust. Accordingly, it is possible to know that organic food is 'certainly' safer than other food.

Yet labellers face an ambiguous relationship with science, with politics penetrating the labelling atmosphere as much as science does (cf. Yearley, 2005). In the case of organic food and agriculture, for instance, green labels have often been introduced with an explicit critique of conventional industrial production and the conventional scientific knowledge production on which it relies. Such knowledge production has been seen as part of the created risks and environmental destruction that motivate actors to demand labels in the first place (cf. Beck, 1992). Several labelling initiators have a far from unconditional trust in science, therefore; rather, their attitude towards scientific knowledge and industrial production is reflective (Boström & Klintman, 2006b).

To further complicate the picture, many other knowledgeable actors claim the right to influence labelling and standardization processes based on their expertise, experience, and practical knowledge. These actors include consumers; professional buyers; marketing actors; social movement organizations (SMOs), including environmental movement organizations (EMOs) and labour unions; business actors, including producers of raw materials, processors, retailers, dealers, and marketing actors; and several types of public officials and individual experts.

All these actors claim the right to influence the processes, not only with their knowledge and experience, but also with their *values*, *interests*, and *political ideologies and visions*. Accordingly, we see it as an important task to shed light on the patterns whereby the tools of green consumerism are created and negotiated within the broad continuum *between* science and politics. Labelling and standards are not based on strictly scientific results; nor are they based on normative and unsubstantial opinions unrelated to facts. This feature places the object of this book in an epistemic dilemma similar to that of many other policy issues that concern the environment and health. We face a difficult challenge in discerning how policies for green consumerism should be understood and treated *vis-à-vis* the epistemic poles of: the objectivist view that information tools of green consumerism provide purely scientific knowledge – a view that we call *epistemic absolutism*; and the relativist view that information tools of green consumerism provide such imperfect information that it is meaningless to favour any green or ethical claim over any other – a view that we call *judgemental relativism*.

In this book we reject both these positions, and aim for a fruitful combination of politics and science, examining how it is played out – or disregarded – in practice. There is a way out of this dilemma: stakeholders and analysts can treat and develop the tools of political consumerism based on an *epistemic relativism*. Our use of this term entails the idea that certain claims (of the criteria used in eco-labelling schemes, for instance), although highly dependent on conflicting viewpoints and 'frames', can be compared and assessed for their effectiveness in reducing problems (Klintman, 2002a).[3]

**Should labels be trusted?**

In policy writings on labelling, the challenges of consumer and producer trust are among the most intensively discussed issues. A Presidential Meeting of the European Union Eco-labelling Board was addressed a few years ago by Michael Ahern, Minister for Trade and Commerce. A major point of his address was that everyone could be assured that 'Manufacturers, service providers and consumers can trust the eco-label [of the EU, the EU flower]' for several reasons, including that it is 'scientifically based', 'supported by public authorities', and 'certified by an independent competent body' (EUEB, 2004). Leaving aside the stated *reasons* for trusting the EU flower, this spokesperson is completely right in his claim that labelling schemes are entirely dependent on public trust (see also Nilsson et al., 2004). We are not usually consciously aware of the pesticides we eat. If we do not want to eat pesticides, we can choose organic options, because organic labellers say that they do not allow pesticides in organic produce, and because they say that they have good auditing and inspection procedures. Why should they cheat? Needless to say, we have little opportunity to gain insight into the numerous circumstances in the production and distribution processes of these organic foods. If we were to know, we might, in fact, consider some of these circumstances to be indefensible: the working conditions for farm workers, for example, and the treatment of animals. Labels are substitutes for our senses and our first-hand knowledge. They provide us with 'mediated transparency' (Klintman & Boström, 2008).

If we knew all the essential details about the circumstances that we want to know, we would not need labels. Labellers have better knowledge about some of the circumstances, or they have good relationships with those who have the better knowledge. We have only to trust – to be convinced – that the labellers and the labels are trustworthy and are doing a good job. But what kind of trust is implied in the relationship between label(ler)s and consumers? Is trust relevant in every case? What type of trust is assumed? Could there be other roles for consumers? This book argues that labellers and stakeholders involved in labelling processes too often wrongly presume *simple trust*. One of the main goals of this book is to develop a new perspective on the relationship between eco-standards and consumers, particularly with regard to the green consumer's concern about and participation and trust in the green tools available. The *simple, unreserved consumer trust in experts*, which is associated with *epistemic absolutism*, is likely to be unproductive, because labelling schemes are not strictly scientific and because they require democratic input about value-based green priorities. In addition, treating green labelling schemes as 'purely scientific knowledge reflectors' is likely to be exposed as incorrect by the reflective public of our late-modern society, which may in turn lead to a *blind public mistrust* in eco-standards, including those parts that are ecologically sound (cf. Power, 1997; Klintman, 2002a). Such blind mistrust is a natural consequence of the opposite epistemic treatment of labelling schemes, namely the *judgemental relativist view* of the standards as completely political and arbitrary. Blind mistrust in the scientific and organizational potential of green labelling schemes and other eco-policies may be democratically and ecologically harmful, because of the consumer passivity, learned helplessness, and cynical reasoning that are often associated with blind mistrust (cf. Zavestoski et al., 2006).

We believe that there are ways to deal with this problematic polarity. Throughout this book we develop the stance that ecologically and democratically effective green consumerism – if that is what is called for – requires that its tools and the policy procedures behind them be designed, modified, and explained in ways that stimulate a third type of trust relationship among consumers and other stakeholders. We call this third type *reflective trust*: a more advanced level of trust, in which consumers and other stakeholders acknowledge the fallibility, ideological diversity, and political compromises of environmental policies. *Reflective trust* is a trust that the standards can be improved, and that consumers and a wide group of stakeholders are needed in these processes of continuous modification – as individuals and as members of organizations.

### Differentiation or integration?

The next theme we developed is based on a polarization found in academic and policy-oriented debates on labelling. Commentators may reject labelling on the basis of two opposite standpoints: one group fearing the marginalization of labelling efforts, and the other fearing mainstreaming tendencies. Constance & Bonanno have investigated the debate following the introduction of the Marine Stewardship Council (the leading global seafood labelling and certification organization), and have made the following comment on this labelling programme:

> TNC interests have … found a novel way to accommodate the 'food movement' demands of green consumers from the North into their global structures of accumulation and legitimation. (Constance & Bonanno, 2000: 135)

This statement represents a view of labelling in which the emancipatory demands of social movements and concerned consumers are co-opted by profit-seeking big business. Too much integration into normal market structures and processes is the problem here (cf. Allen & Kovach, 2000; Raynolds, 2000; Guthman, 2004).

In contrast, other commentators maintain that labelling is likely to remain inefficient because it will cover only a marginal market share. The most common argument is the (uncertain) prediction that consumers' willingness to pay extra for eco-labelled products will remain modest (Batte et al., 2007). Such arguments are common with regard to alternative energy sources:

> Thus, households often express a strong support for 'green' electricity but these attitudes are more seldom reflected in active choices of 'green' power suppliers. (Ek, 2005: 1677–1689; this author refers to previous research rather than giving her own view)

Pessimists often argue that we should not expect labelling to be powerful or influential because labelled products will always have only a marginal share of the market. Most products will remain conventional, cheap, and dirty. Green products, the argument goes, will continue to constitute a niche market, and that niche will remain a small one. People will continue to be free-riding ('Why should I pay more for labelled products when other consumers don't do so?'). The pessimists also say that labelling could be more powerful if public authorities enforced durable and extensive implementation (an alternative that creates its own problems, such as violation of free-trade rules). They say that labelling is unrealistically dependent on a stable public opinion with a significant share of 'green political consumers'. Although many people express their willingness, concerns fluctuate, as does market demand. Hence, in these arguments, niche appears as the problem.

On the other hand, one could argue that niche appears simultaneously as part of the solution, the reason behind the power of labelling. An argument we promote in this book is the notion that labelling is able to obtain its power because of, rather than in spite of, its image as a niche. This has to do with an essential characteristic of green labelling, which we have many reasons to return to in this book: namely that the labelling is based on *symbolic differentiation*.

## Do the labels accord with the concerned consumers' concerns?

Our final working hypothesis has been that there is a discrepancy between what is presented to consumers on the front stage (through categorical and overly simplistic ecological messages), and what is actually taking place on the back stage (where the eco-standards are created and negotiated). This book discusses whether and how such a gap appears and the ecological and democratic advantages of bridging the gap.

Our empirical focus suggests that we have more to say about the production side than the consumption side, yet the literature on political consumerism, along with related literatures, is useful for finding indications of consumers' concerns, which is, in turn, useful for analysing and discussing the relationship between production and consumption.

Political consumerism refers to the idea that many late-modern consumers express non-economic values (e.g., concerning human rights, animal rights, global solidarity, and environmental responsibility) through the market arena. Political consumerism may, for instance, be conducted through boycotting or buycotting (Micheletti, 2003; Micheletti et al., 2004; Sørensen, 2004; Boström et al., 2005; cf. Harrison et al., 2005; Klintman & Boström, 2006; Zaccai, 2007). The literature indicates a positive trend, in that an increasing number of people consciously boycott or buycott products for political and ethical reasons.

Political consumerism can be seen as an example of 'individualistic collective action', in contrast to 'collectivist collective action' (Micheletti, 2003), which, in turn, refers to traditional patterns of political participation within nation-state representative democratic structures. The latter are 'frequently viewed as time-consuming, limiting in terms of individual expression, and lacking a sense of urgency' (2003: 24). Rather, people look for more flexible, spontaneous, everyday channels to express engagement and responsibility about various issues. Political parties are seen to be inert and as having difficulty integrating new issues into their ideologies and actions. People express their political views and identities by choosing green mutual funds, organic food, and fair-trade-labelled clothing; they make political statements by visiting localities that feature eco-tourism; and some people – many vegans and vegetarians, for instance – play their part in the reforming of modern food production by avoiding entire product categories. Thus, labels seem to accord with general trends towards political individualization and seem to fit the agendas of contemporary concerned consumers.

To be sure, the literature does not endorse this trend uncritically. One stream investigates the consumer groups that are and are not active in this type of political participation, and discusses the democratic consequences of such biases (see Chapter 4). Although we acknowledge the value of these studies, we want to explore another problematic side of political consumerism: the relationship between the *production side* and the *consumption side* of the labels. The extensive studies say little about consumers' thoughts, assumptions, and reflections about the tools. The researchers appear to assume that eco-labels, fair trade labels, or similar arrangements resonate with the identities, hopes, and political sentiments of consumers. Based on the view that we develop in Chapter 4 of the 'typical' consumer; that is, reflective, ambivalent, and uncertain consumer, capable of developing reflective trust, we ask: Is it really possible to reach people with simple and unambiguous tools, with adequate information? Do policymakers within labelling activities develop a simplistic view of consumers – that they are overly autonomous and rational, for example (cf. Cohen & Murphy, 2001)? If there is a mismatch between labelling tools and consumer trust, is there a threat also to the long-term sustainability of the labelling instrument itself? How can green labels and other consumer-oriented tools be developed that better match the ambivalence and reflective potential among concerned consumers? These are indeed big questions, which we will elaborate upon. In Chapter 4, we try to provide certain patterns indicative of political consumerism, and in Chapter 11 we provide ideas for future research on how to reduce any mismatch between concerned political consumers and consumer-oriented tools.

## Analysing labelling: outline

In the concluding chapter, we briefly return to our four themes: politics, trust, differentiation, and mismatch. Our ambition, which requires us to go through a number of steps, is to be in a position to develop these four related arguments. To begin with, we consider it informative to relate the emergence of labelling to a historical context, including five key trends. Chapter 2 relates labelling to such key trends as individualization, globalization, ecological modernization, the shift from production to consumption, the shift from government to governance, and the rise of private authorities and new rule-making. This historical overview is followed by an exploration of what labelling 'really is', given the spectrum of emblems, seals, badges, and signs in society and on the market. Consequently, in Chapter 3 we say a few words about the variety of eco-standards that have been introduced during the last decades. The chapter also provides a definition of green labelling and of eco-standard.

Chapter 4 turns to the end consumers and to the phenomenon of green political consumerism. There is a growing body of literature on today's more concerned consumerism, and on 'ethical', 'political', and 'green' consumers. The chapter begins by giving a brief and selected survey of this literature, leading to an elaboration of typical concerned consumers and the challenges they face. The chapter introduces a number of possible goals of policy tools to encourage green consumerism, by distinguishing among consumer insight, trust, and influence.

Chapter 5 introduces our cases: organic labelling, forest certification, GM labelling, seafood labelling, green and ethical funds, green electricity, and paper labelling. Here we present basic information about the cases, and discuss reasons why these cases have been selected.

In Chapter 6 we use a method that is unusual in writings on policy research. The aim of the chapter is to remove the subjects from the main arguments about green labelling – sceptical arguments as well as encouraging ones. Thus, the chapter surveys these arguments without discussing or interpreting the groups of actors – the industry, EMOs, consumer organizations, governments – that typically make these arguments. The idea behind this faceless overview is to present a toolbox of arguments, in order to demonstrate in subsequent chapters the flexibility that various actors exhibit in choosing their arguments. The critical and endorsing arguments are categorized by themes, oriented towards the market, knowledge, and green governance. In addition to reading it through, the reader may use the chapter encyclopaedically.

The next four analytical chapters present a mixture of theory and empirical data. In our analysis of labelling, we broadly distinguish between policy context and process factors (framing and organizing). In Chapter 7, we discuss our understanding of the relationship between context and process factors (Figure 7.1: Policy contexts and labelling). Because our book is embedded within a political sociological tradition, the role of the *policy context* is critical, as shown in this chapter. We analyse how green labelling processes relate to certain globalizing tendencies and transnational rules; yet we also attend to the development of eco-standards and their dependence upon place and history. Our cases illustrate how traditional regulatory bodies – along with political cultures, existing regulation, the organizational landscape, and the infrastructure and materiality of the product and production processes – play an important part in our understanding of the development of new policies.

Our analysis of labelling processes is further influenced by what has been called *interpretive policy analysis*, in which the so-called *frame analysis* is useful for understanding debates, discussions, and compromises in green labelling and standard setting. Green labelling is a process in which uncertain, dispersed, and complex knowledge, as well as diverging values and interests, are translated into a simple and categorical label. Through this process of categorization, the people and organizations involved include certain limited and manageable parts of reality, whereas these labelling actors either overlook or exclude other parts of reality as 'irrelevant', 'extreme', or 'unfeasible'. Although we use the framing perspective throughout the book, Chapter 8 provides a systematic analysis of three underlying strategies within the framing process: *boundary framing*, *frame resolution*, and *frame reflection*. These framing strategies are partly related and sometimes overlapping, but they may also conflict with each other. The three framing strategies create exclusion and inclusion of substantial environmental themes. One intriguing question concerns what is – or should be – included and excluded (substance), whereas another question concerns the nature of the procedure: whether, and to what extent, an open *frame reflection* is part of the process across the people and groups involved, for instance.

Finally, our study employs an organizational perspective on standard-setting processes. Labelling always occurs in an organizational context, which includes social movement mobilization and coalition-building for or against labelling. The development of organizational forms for the labelling may have great impact on the possibility of engaging in constructive dialogue, reflections, and cooperation among actors, including consumers. Chapter 9 analyses the various explicit and implicit tasks a labelling organization sets out to perform, and the roles of a number of key participating actors. Common in labelling – both in the initial phases and in subsequent standard-setting processes – is an intense interaction between business and social movements. But *how* such interaction occurs may vary substantially. We investigate three types of interaction in labelling arrangements: business-governed labelling (in which SMOs may assume outsider or advisory roles); hybrid-governed labelling (in which business actors and SMOs negotiate and share decision-making power); and SMO-governed labelling (in which business actors play advisory roles). How does such variation affect the formulation of labelling principles and criteria, as well as debates, framings, and power struggles among groups?

One of the key organizational challenges in labelling is how to deal with the mutual mistrust among the actors that appear to be important to include within a labelling arrangement. To develop standard criteria may require that groups representing different expertise and interests initiate a dialogue, and that controversies and disagreements are resolved. They may lack experience with such dialogues, however, may not be keen on communicating on equal ground, and may believe that others are pushing hidden agendas. In Chapter 10, we investigate why and how groups in some of our cases have managed to develop mutual and reflective trust, in spite of their initial mistrust, whereas such rapprochement has failed in other cases. These analyses focus on the role of repeated interaction and cognitive authorities, and on the roles of transparency and auditability.

In the concluding Chapter 11, we discuss our general findings in relation to the four themes presented in this chapter, and connect the themes to urgent policy issues of today and the future. In this chapter it will be clear to the reader that we allow ourselves to engage in the issue normatively.

Before moving to the next chapter, we take a step back and discuss the methodological choices and the normative positions behind this book.

## Methodology and normative position

The empirical basis of this book is the extended, comparative, case-study approach. By 'extended' we refer to its methodological pluralism, with the use of several types of data sources (see below). We have compared labelling projects in two *countries* (Sweden and the United States), across several *sectors* (forestry, paper products, fishery, organic production and food products, GM food, green/ethical funds, and green electricity). The primary aim of our selection is that the cases should cover and analyse a wide range of challenges. The cases have been selected to allow for the analyses of different types of framings (such as precaution-ary framings in Europe/Sweden vs 'yes, unless' framings in the United States), organizational arrangements (SMO-governed labelling, Hybrid-governed labelling, and Business-governed labelling), policy contexts (e.g., the role of adversarial vs consensual political cultures). Although it would probably be possible to replace a few of our cases with others, we believe that our selection has enabled us to shed light on breadth, which was our most important rationale for case selection. It should be stressed that the main objective of using data from Sweden and the United States (along with secondary data from other European coun-tries) is not to give finite answers to national similarities and differences in green consumer policies. Rather, the aim is to focus on challenges and opportunities for green labelling and consumer policies, as pre-sented in our four themes. The countries and sectors we focus on serve as examples rather than as final research objects in our book.

Although all our case studies have, roughly speaking, been accorded similar time and attention, there are certain variations and imbalances between the different cases in terms of our primary research. We did not strive for complete symmetry in the research design of the various case studies. Rather, such a strategy would have been counterproductive, because it has been apparent to us that some of the cases shed light on particular aspects of labelling practices. Consequently, systematic comparisons along the same dimension have not always been the primary methodological goal. In Chapter 5 we discuss the rationale behind the selection of each case study in greater detail. The case studies on forestry certification, organic food labelling, and green mutual funds refer to both the United States and the European/Swedish situations. The case studies on fishery and paper products concern the Swedish setting and the case study on GM labelling refers mostly to the US situation.[4]

A degree of symmetry is, of course, indispensable, to the extent that one wants to explain similarities and differences – as is also our ambition. Indeed, our decision to compare labelling processes across both countries and sectors offers multiple comparative opportunities. Identifying similar types of dilemmas in different countries or sectors, for instance, may be an indication of a general pattern. The reader will notice that we do not compare every case for every new topic we address; rather we pick the case that best illustrates a point we wish to make. Yet all cases appear implicitly in all our analyses, so our narrative approach should not be confused with anecdotal evidence-making.

We maintain that the United States and North-Western Europe present contrasting and illustrative settings. For our comparative purposes, the two regions represent a fascinating difference in the commonly observed consensual policy climate of North-Western Europe, which many political analysts claim is in stark contrast to an adversarial, conflict-oriented North American policy climate.[5] The intensive case-study approach can be extremely useful for contrasting one setting with an opposite one, and for providing deeper understanding of the specific conditions for policymaking in each setting (cf. Christensen & Peters, 1999; Blyth, 2002). Intriguingly, however, our secondary data from other Northern European countries indicate, in certain respects, results that are highly different from the Swedish ones – something that is discussed later in this book.

Another way to enhance the potential of comparisons is to compare different projects in the same country. We expected that this strategy would, among other things, demonstrate that one singular national context does not determine how labelling projects are conducted. It is also important to examine traditions in different sectors and to emphasize the process-oriented factors. In several countries, for instance, the electricity sector (in which priorities of various energy sources and labelling criteria have been debated) has been subject to far more conflict-impregnated debates and policy processes than have many other sectors that are also present in traditionally consensus-oriented countries.

For each case study, we have used documents such as websites, reports, minutes, newsletters, stakeholder comments on draft standards (Swedish *remisser*), press releases, and not least surveys, as well as other research conducted by the labelling organizations. We have also interviewed key persons representing various key organizations participating in labelling projects (social movement organizations, scientists, authorities, business, labour, and labelling administration).[6] Furthermore, we have used secondary sources. Although we have not followed the processes in real time, this combination of methods and sources has enabled us to employ a certain longitudinal perspective, by providing the analysis with a historical and dynamic dimension.[7]

Where can our analysis of data across countries fit within the intricate phenomenon of globalization? Green labelling is indeed part of – and affected by – various 'global flows' (Oosterver, 2005). This longitudinal perspective allows us to account for the global dimension in our analysis, along with the use of various secondary sources. Examinations of these broader data across time and national borders have helped us to avoid or to see the limits of static, cross-national comparisons.

A few final words should be raised about the normative position underlying this book. It goes beyond our aim to prescribe to policy actors the concrete decisions they should make. Nevertheless, we do intend to recommend what aspects and challenges of consumer policies they should be considering. As to labelling, we start with the presumption that such consumer-oriented market instruments exist and that they are seen by various groups of actors as promising tools for meeting economic, social, environmental, and health-related challenges. Hence, whereas the book mentions the more fundamental issue of the consumed *amounts* of products, our focus is on the improvement of labelling and other market-oriented eco-standards, in terms of consumer involvement in setting the criteria for the standards and in terms of less harmful outcomes. Still, one cannot avoid the possibility that there is another type of 'improvement' of such consumer instruments: that they should not suffocate or silence more fundamental calls for reduced amounts of consumption.

# 2
# The Historical Context – Key Trends

This chapter situates labelling, and its concrete manifestations, within a broader historical context, and briefly considers how labelling relates to broad general trends.

> Debating the direction of consumer activism is not new, nor is the attempt to organise disparate individual acts of consumption by appealing to higher moral or political ends. The US-nonimportation movement of 1764–76 was America's first consumer revolt. Aimed against the import of goods, it was more than a rejection of colonial tax laws, /...namely.../ an expression of cultural independence and an assertion of the local over the global. (Lang & Gabriel, 2005, p. 40)

Consumer movements have since this consumer revolt organized themselves into a number of different streams, on both sides of the Atlantic (Lang & Gabriel, 2005; Soper & Trentmann, 2007). Some, such as the cooperative movements that emerged strongly in the United Kingdom and elsewhere in Europe from the mid-nineteenth century, connected with socialist movements, encouraged 'self-help by the people', and expressed radical visions for societal change (although we see extremely little of this radical spirit in these days' large-scale business and retail-led coop 'movements'). Other less radical initiatives were mainly taken in the United States in the late nineteenth and early twentieth centuries. Groups such as the National Consumers League and Consumer Research Inc. (later renamed Consumer Union) took form with the aim of informing and educating consumers to help them achieve value for money. This movement eventually spread to all corners of Europe, either organized into large consumer associations as in the United Kingdom (The Consumers' Association, which issues the well-known magazine *Which?*) or in the form of consumer agencies, such as the Consumer Agency in Sweden. An important task for such associations and agencies was, and is, product testing and the provision of research, information, and recommendation to consumers. Particularly the American scene has also seen a more radical variant of this stream, namely 'Naderism' (see Lang & Gabriel, 2005). Ralph Nader has encouraged a more active involvement of consumers and citizens, an involvement that goes beyond the use of package information and product labelling. An important theme has been to foster a deeper understanding on how powerful large corporations may easily dupe individual consumers.

Nor is *labelling* a completely new phenomenon. Already in the late nineteenth century the National Consumers League, in their White Label Campaign (1898–1919), undertook an initiative for 'fair trade'-labelling cotton underwear (Sklar, 1998; Micheletti, 2003). However, this is rather to be seen as an exception to the rule that consumer history generally lacks labelling initiatives. Micheletti's (2003) historical overview of 'positive' (buycotting) or 'negative' (boycotting) political consumerism gives few indications of labelling initiatives in the past (yet also with biodynamic agriculture as an important exception) and plenty of examples of boycotting strategies, especially on the American political scene.

Green labelling should be seen as a significant part of a rather recent trend in environmental politics and policymaking, which goes hand in hand with an increasing prominence for positive rather than negative political consumerism in both Europe and North America. Earlier consumer movements and activism, such as those mentioned above, prepared the ground for new environmentalist initiatives. Important agenda-setting took place during the 1970s and 1980s with key players such as the International Federation of Organic Agriculture Movements (IFOAM) in the organic case and the German Blue Angel in other everyday products. By the early twenty-first century, several dozen national and international eco-labelling schemes were set up around the world (Berry & McEachern, 2005). In addition, programmes for organic labelling, forest certification, fair trade labelling, and green mutual funds are developing fast and worldwide.

Before looking more closely at such labelling arrangements and their definition in the next chapter, we will discuss a few general historical trends that pave the way for a phenomenon such as labelling. We are not trying to provide an exhaustive list of exogenous factors that explain the rise of labelling, but wish to draw attention to five key trends: (1) individualization; (2) globalization; (3) ecological modernization; (4) a shift of orientation from production to consumption; (5) a shift from government to governance, the rise of private authorities, and new rule-making.

## Individualization

Individualization indeed has a long history. Classical social theorists, such as Karl Marx, Emile Durkheim, Max Weber, Georg Simmel, and Ferdinand Tönnies, discussed individualization, that is, how modern urban life makes people more self-dependent and detached from traditional social bonds and communities. This is the case whether the issue is to provide industry with capable and competent (alienated) labour or to choose one's own way of living and believing. Contemporary social scientists such as Anthony Giddens and Ulrich Beck repeatedly stress how such tendencies are radicalized in late-modern life due to both negative and positive developments.

On the one hand, economic growth, rising welfare, extension of education, and popularization of science give late-modern man more freedom, resources, knowledge, and reflective capacity than ever before. People are decreasingly dependent on given traditions, cognitive authorities, class positions, or gender belonging. Identity and lifestyle are increasingly subject to self-reflection and choice. In the political domain, we see a move from 'collectivist collective action' towards 'individualistic collective action' (Micheletti, 2003).

On the other hand, it is important to emphasize – as Beck and Giddens do – that the new freedom of market liberalism has not succeeded in making everyone happier. The emergence of high-consequence, irreversible risks forces people to reconsider 'truths', values, habits, norms, and lifestyles. Confronted with a multitude of late-modern risks and regulatory failures, people cannot choose not to choose. Confidence in old traditions and authorities turns to mistrust. The individual is forced to behave and think as an 'individual', to assume responsibility for actions performed or not performed, and to judge expert advice and counter-advice. The unintended risks created by welfare society, and the systematic failure of existing regulatory and scientific institutions to anticipate these risks, mean that people must deal with the risks themselves.

'Detraditionalization' (Giddens, 1990) implies dissolving given recipes for how to live and how to choose, meaning that individuals need new codes, signs, or recipes for this. Some opt for ignorance or cynical reasoning whereas others choose reflection and changed habits.

## Globalization

Globalization is today's key term to interpret all sorts of current societal processes. We may relate green labelling to a number of globalization processes: economic, financial, political, ecological, technological, or cultural. For example, labelling could be seen as a promising regulatory response to the increasing inability of nation states to regulate transboundary risks (Beck, 2000; Oosterver, 2005; see also below 'from government to governance'). From a cultural perspective, labelling connects with global flows of symbols and messages, and with the world culture of virtue, voluntarism, and rationalization (Boli & Thomas, 1999).

There is also an interesting affinity between the concept of globalization and the concept of labelling. Recall that globalization is about increasingly complex forms of interaction and organization on various scales. Globalization is about the reorganization, compression, of time and space (Giddens, 1990; Held et al., 1999). Globalization is about the distant becoming closer, familiar, and local. The opposite is equally true: globalization is about the close, familiar, and local becoming distant. Labelling too is about complexity and distance. Labelling involves translating something complex, abstract, and distant into something close, concrete, visual, and familiar, while at the same time preserving a certain degree of abstraction and distance. A label on a product package is something concrete and abstract at the same time.

## Ecological modernization

If we move to a less abstract level, we may trace histories of labelling in relation to the history of environmental protest and movements. A first wave of environmentalism emerged some time in the late nineteenth century when a concern for nature protection arose in the United States and Europe (van Koppen & Markham, 2007). From such a historical perspective, strong attention towards green consumerism and labelling is a recent environmentalist strategy. The 1960s and 1970s saw the radicalization of environmental protest and the emergence of ideas such as 'limits to growth' and 'small is beautiful' (Hajer, 1995). 'Alternative lifestyle' emerged as a key concern, which is a basic message that contrasts with today's environmentalism. Alternative movements of the 1970s centred on the imperative 'Consume Less'. These movements constructed framings around over-consumption, self-sufficiency, green communes, reliance on local natural resources, smallness, holism, and localness.

However, this alternative movement remained marginal, appeared backward-looking, inward-looking, extreme, romantic, and even ridiculous to many audiences. Hence, the new environmentalists of the early 1980s were not entirely happy with previous approaches (Hajer, 1995). They saw few reasons to continue calling for radical change and fundamental reorganization of the social order. Yet, chemical risks, nuclear risks, food scandals, and other risks gradually raised new concerns among the public. In addition, the 1980s and 1990s saw a gradual rapprochement among environmental movement intellectuals, business elites, and policymaking elites. Ecological modernization developed as a dominant new approach for conceptualizing environmental problems, with a fundamental belief in progress and the problem-solving capacity of modern techniques and institutions. From the perspective of the ecological modernization discourse, environmental problems can be calculated, solved, and anticipated. That is possible without altering the foundations on which modern institutions rest. New policymakers framed the win-win scenario between economic and environmental development, and the emerging sustainability discourse extended this line of thinking by adding a third pillar – social development. From now on: 'A radical farewell has been said to the small is beautiful ideology, and technological developments were now seen as potentially useful in regulating environmental problems' (Spaargaren & Mol, 1997, p. 84). Its new focus was not on abolishing capitalism or industrialism altogether but on reforming concrete economic activities. Cooperative strategies, win-win arguments, proposals for practical solutions, and the symbolic demonstrations of 'good examples' were a few of the new tactics used by environmentalists (Boström, 2004a). Green labelling was a promising new strategy that accorded with this general political, ideological, and discursive shift.

## From production to consumption

In parallel to the above-mentioned trends we see a general shift from production to consumption in all corners of political and social life. In the United States, consumption has historically been higher on the political agenda than in Europe. Because the socialist workers' movement was never strong in the United States, labour unions and other movements, such as the civil rights movements, were forced to use consumption – organized boycotting – as a strategy for their political struggles (cf. Strasser et al., 1998; Micheletti, 2003; Vogel, 2004). In her study of the US consumer movement, Lizabeth Cohen claims that The

Great Depression of the 1930s gave rise to the American 'citizen-consumer'. These citizen-consumers (i.e., the general public) regarded themselves – and were regarded by policymakers – as guarding the rights and safety of individual consumers. Moreover, they were viewed as exercising purchasing power, thereby contributing politically to the larger society as well (Cohen, 2003, pp. 18–19). Another example is ethnic conflicts, such as the 'Don't Buy Where You Can't Work' campaign, which started in the 1930s to obtain white-collar jobs for African American residents (Greenberg, 2004).

By contrast, for labour unions and many other groups in Europe, consumption has mostly been considered secondary to production. 'Europeans have tended to view the boycott weapon as a political tool that is more difficult to control and therefore less appealing than other more organized forms of struggle' (Micheletti, 2003, p. 46). Although there is no lack of historical examples of consumer regulation, movements and activism on both sides of the Atlantic, which we mentioned in the introduction of this chapter, production has received far more attention everywhere. However, the focus on consumption – particularly positive political consumerism (boycotting) – is increasing dramatically in both the United States and Europe.

Consumption re-entered the environmentalist agenda in the 1980s. In contrast with the previous, profoundly American, negative political consumerism (boycotting) and compared with the Consume Less radicalism in the seventies, it was a new type of consumer focus, which resembles the ecological modernization paradigm. All types of actors developed and embraced the new approach to consumption, which some scholars called 'modified consumerism' (criticized by Lintott, 1998, p. 240) or 'reflexive consumerism' (Macnaghten & Urry, 1998, p. 25) or just 'green consumerism' (e.g., Lang & Gabriel, 2005). EMOs actively sought to disseminate a new view on lifestyles. Rather than providing alternative lifestyles and complete recipes for how to design one's life, EMOs now propagated that people could integrate environmental messages into their existing lifestyles (Boström, 2001). In their already classic book from 1990, *The Green Consumer Guide*, Elkington, Hailes & Makower encouraged consumers to use their own environmentally friendly potential: 'By choosing carefully, you can have a positive impact on the environment without significantly compromising your way of life. That's what being a Green Consumer is all about' (Elkington et al., 1990, p. 5).

Rather than giving moralist lectures to the public about the need for a radical transformation of modern life, which were said to turn the broader public off in the 1970s, environmental groups are now disseminating concrete recommendations that do not threaten the comfort of prevailing modern lifestyles. From now on consumers are part of the solution rather than simply part of the problem, and they can do something proactively without completely altering current habits and interests.

The proliferation of EMO-supported shopping guides in the 1980s was a direct precursor of eco-labelling schemes. These shopping guides offered positive (what to buy) or negative (what not to buy) information (Berry & McEachern, 2005). Consumers were familiar and therefore receptive to such types of advice because well-known consumer organizations in different countries already had their routines and magazines for informing about 'best buys' and 'bad buys' (Lang & Gabriel, 2005).

A problem with green shopping guides, however, is the uncertainty regarding the quality and credibility of the information. Debaters expressed similar worries over the greenwashing of businesses, made possible through the rise of many green trademarks in the same period. Who collects and sends such information, what kinds of expertise and interests do they represent, and on what criteria are companies or products assessed? Such concerns, but also positive initial experiences with strong demand for informal consumer guides, spurred rationalizing efforts to establish labelling arrangements.[8]

The 1990s saw a generally increasing focus on consumption in Europe. The environmental movement was greatly helped by all the European food scandals in the 1990s, beginning with the BSE crisis (Carson, 2004), in mobilizing a huge, powerful ally: a risk-conscious and anxious public. Suddenly, many more people appeared willing to use their own pockets to favour or disfavour certain products or producers.

The shift from production to consumption has contributed to the increasing focus on improving corporate accountability and transparency during the last 15 years (Adams & Zutshi, 2005; Crane, 2005). A concrete sign of this is the initiation and development of a whole range of codes and techniques aimed at stimulating and visualizing responsible corporate conduct (Boström & Garsten, 2008; and see Chapter 3 on eco-standards). The use of symbols, signs, and entire brands for communicating the greenness of companies grows steadily (Peattie & Crane, 2005). As O'Rourke maintains, 'brands [and trade marks] have become in many ways the primary currency – and central piece of information needed – for global sales of products' (O'Rourke, 2005, p. 119). The most well-known companies with early success in giving their trademarks a green image include The Body Shop, Ben & Jerry's Homemade ice cream, and Patagonia (see, e g , Weinberg, 1998).

As the importance of green branding increases, the vulnerability and visibility of the corporate brand increase in the marketplace; and 'many brands are facing something of a trust deficit in terms of the public's faith in their commitment to "doing the right thing"' (Crane, 2005, p. 227). A challenge for green advertising is that many green consumers have anti-corporate attitudes (Zinkhan & Carlson, 1995). Typically, they also mistrust the advertising industry, and are inclined to see through vague and over-ambitious green advertising claims. Increasingly, companies learn they have to manage a new situation with more concerned consumers and with new risk of negative publicity and social movement campaigning (Crane 2000, 2005; Holzer, 2006, 2007; Power, 2007). Not merely image and impression management, but also 'reputational risk management', appears as a key corporate activity, and the issues of greenwashing and not 'walking the talk' become key topics in debates surrounding responsible corporate behaviour. We shall see later that green labelling is a response to such challenges in green branding, marketing and advertisements.

Along with this general trend from production to consumption, a trend which is visible in state and market arenas, a new type of scholarly literature has emerged. Scholars have begun to rethink the 'citizen' – 'consumer' divide (Soper & Trentmann, 2007); the body of literature on political, ethical, and green consumerism is growing (see Chapter 4).

## From government to governance, the rise of private authorities, and new rule-making

Whereas the lion's share of the emerging literature on green consumerism focuses on the demand side, on consumer frontstage problematics, this book focuses on the supply side of labelling tools. It is accordingly necessary to take into account institutional policy arrangements behind labelling, including the political, organizational, and discursive (framing) structures and processes. In general, the rise of such policy arrangements as labelling corresponds with general types of institutional trends such as 'from government to governance' (e.g., Rhodes, 1997; Pierre & Peters, 2000), or the rise of 'private authorities' (e.g., Cutler et al., 1999; Hall & Biersteker, 2002; Rosenau, 2003).

Among both policymaking and academic circles, a general concern with old-fashioned command-and-control type of regulation grew. Throughout the 1980s and 1990s a consensus emerged that top-down, *ad hoc*, and reactive 'end-of-pipe' strategies should be avoided or at least supplemented by more anticipatory, cooperative, and systematic ways of regulation. Again, this follows the 'eco-modern' way of thinking. A turn towards more inclusive and cooperative forms of policymaking grew (Lafferty & Meadowcroft, 1996; Mol et al., 2000; van Tatenhove et al., 2000). The old mode of regulation, it was argued, was unable to deal with diffuse, complex, large-scale, and transboundary environmental risks and problems (cf. Giddens, 1990; Beck, 1992).

Regulatory failures, legitimacy crises, and public mistrust of existing regimes stimulated all types of policymakers, in both state and non-state arenas, to look for new types of instruments and arrangements. What we see in the environmental field is a huge number of regulatory innovations, called 'joint environmental policymaking' (Mol et al., 2000), 'multi-stakeholder dialogue' (Bendell, 2000), 'partnership' (Glasbergen et al., 2007), 'informational governance' (van den Burg, 2006), and so on. A common trait among the approaches is that they rely on voluntariness and consensus-orientation (Boström & Garsten, 2008). Those to be regulated are seen as knowledgeable, responsible, and capable actors.

Some scholars speak of a third wave of environmental policies which includes green labelling, along with command-and-control regulation and economic instruments (taxes, subsidies). Environmental textbooks often label new instruments of this kind 'voluntary approaches', such as 'information', 'education', and 'environmental management systems' (e.g., Connelly & Smith, 2003).

Voluntary approaches are a family of policy instruments that fit general trends such as individualization and ecological modernization. If organizations and people see themselves as free and independent, they can be expected to be more receptive to information, which they are free to use according to their own interest. Is this not purely a case of neoliberal hegemonic ideology? The focus on market-based approaches and voluntariness has certain similarities to such thinking. It is definitely no accident that the emergence of labelling coincides with a strong global ideological movement that stresses the role of markets not only in *creating welfare*, but also in *solving problems with the side effects* (externalities) of economic development.

On the other hand, it is important to stress that such instruments as labelling, despite being voluntary, require market intervention. The energetic call for deregulation in the 1980s and early 1990s led in many circumstances to regulatory void and, as a consequence, to calls for re-regulation (through soft or hard approaches, or a combination of both). The years since then have therefore seen a virtual explosion of regulatory innovation and new rules in all areas of political and social life

(Brunsson & Jacobsson, 2000; Ahrne & Brunsson, 2004a; Mörth, 2004; Djelic & Sahlin-Andersson, 2006; Power, 2007; Boström & Garsten, 2008). There is a demand for rules also among those who are supposed to comply with them.[9] For example, in Chapter 6 we will present some market-oriented encouraging arguments in relation to green labelling, which indicate a strong demand for rules.

One type of regulation in particular has caught our attention, namely standardization. The concept of standardization helps us to define green labelling and to better grasp some of its characteristics and dynamics, and this will be the main topic in the next chapter.

# 3
# Green Labels and Other Eco-Standards: A Definition

Brunsson and Jacobsson (2000) argue that *standardization* emerges as a general new form of regulation in modern globalized life, alongside traditional legislation and normative community. The conventional understanding of standardization is that it deals with technical objects or systems (nuts and bolts). However, standardization increasingly extends to include social and environmental matters (Busch, 2000; Cochoy, 2004; Tamm Hallström, 2004, 2008). All types of activities among organizations and individuals can be subject to standardization. What should an organization look like, what should be its aspirations, what types of administrative routines should it have? How should we design an education programme? What should we eat? What are our rights and duties? There are thousands of concerns in everyday life that could be subject to rule-making, but for which the alternatives – traditional legislation and normative community – appear inadequate.

Green labelling can be seen as a policy instrument or as a particular kind of information (Jordan et al., 2003; van den Burg, 2006). These are both adequate views, because labelling is about steering actors (policy instruments) and informing about buying options (information). However, we think it is essential also to see green labels as a kind of eco-standard. A standard is a kind of *rule* – or is made up of a family of rules such as 'principles' and 'criteria' – next to other kinds of rules. Following Brunsson and Jacobsson (2000), we see standards as *voluntary rules* in contrast to directives (such as law). And standards are explicit, written, codified, in contrast to norms. Norms are implicit rules that are often taken for granted, and which enable interaction in normative communities. Standards can have various contents, however, referring to both substantive and procedural matters. They can be abstract or concrete, precise or vague. Furthermore, the very same rule-content,

27

such as 'do not throw rubbish into the sea', can simultaneously be a norm (it is commonly assumed that this is bad behaviour), a codified standard (e.g., part of a CSR programme), and a directive (e.g., part of a statute).[10]

*Eco*-standards are standards that are addressed towards solving or dealing with environmental problems (whereas economic and social aspects could be incorporated as well). Next, we provide a definition of green labels and thereafter briefly discuss other types of eco-standards.

## Defining green labelling

Conceptually, we prefer to use the term *green labelling* rather than the empirical term *eco-labelling* (although we use the term eco-label in cases where it is relevant) because the former is more general in that it covers related tools, such as *stewardship certificates, green mutual funds,* and also some *green trademarks*. We define green labelling in the following way:

> As a kind of eco-standardization, green labelling is based on the standardization of principles and prescriptive criteria. This type of eco-standard is market-based and consumer-oriented, and it relies on symbolic differentiation.

Green labelling is based on standardization of principles and prescriptive criteria. Producers who want to use a label on their products must comply with these standards and normally pay a licence fee. Labelling criteria are normally – but not necessarily – set by a party that is independent of the producer. Plenty of, but not all, green labelling schemes are based on independent third-party certification – an external auditor examines whether the certified company properly complies with the standard and has the authority to require corrective measures and, in the event of continued non-compliance, to withdraw the certificate. Criteria and principles are not fixed in stone. In contrast, many labelling programmes express a vision of adjusting, developing, and sharpening labelling principles and criteria in a continuous manner, in the light of new knowledge and market opportunities.

Green labelling is, moreover, a particular kind of market-based instrument, in contrast with other instruments that also rely on market dynamics (e.g., taxes, subsidies, quotas, and emission permits). Green labelling is a market-based *and* consumer-oriented approach to dealing with various environmental issues. Green labels are markers which are presented to consumers or professional buyers, and which symbolize beneficial consumer choices in terms of environmental, health, quality, solidarity or other matters. Compared with economic instruments such as taxes and subsidies, green labelling requires a demand for green labels among endconsumers and/or purchasing companies. It should be emphasized, however, that an 'imagined' or 'represented' demand could be at least as important as a 'real' demand (cf. van den Burg, 2006). To be sure, the existence of conscious, ethically and politically motivated consumers or professional buyers who voice their willingness to express their values and interests through green products (and often to pay a bit extra) by means of the market channel is an essential vehicle for green labelling. However, political consumers are in general an individualized and disorganized category of actor. EMOs often play essential roles in mobilizing, empowering, demonstrating, and aggregating this kind of latent disorganized consumer power *vis-à-vis* business and other audiences (Gulbrandsen, 2006; Holzer, 2006).

Finally, green labelling essentially relies on *symbolic differentiation*. A green label is a symbol. The labelled product is unable in itself, by its sheer visual appearance, to show whatever it is that someone wants shown. This could be something inherent, but invisible, in the product or something integral in the production process behind the product. In addition, the label symbolizes that a particular product has a quality – in a positive or negative sense – that equivalent products (or substitutes) lack. The symbol says implicitly that this product is *different* from other products, often discursively signalled as 'conventional products'. We will argue several times in this book that this kind of symbolic differentiation is a key to understanding dynamics in labelling, including such processes as identity construction, marketing, positioning, and scheme development. At the same time, we will also have reason to get back to a counter-dynamic in that green labelling also has to fit in and be integrated with existing market and industry structures. This double need for differentiation and integration is likely to cause tensions and contradictions in labelling processes.

We will introduce our green labelling cases in Chapter 5, but it is relevant here to mention a few prominent examples. The first fully developed nationwide eco-labelling scheme was elaborated in Germany in 1978, the well-known 'Blue Angel' (nicknamed after the UN's logo). It spurred the development of schemes in many other countries, although debaters disagree as to how successful – and successful in what sense – it has been (see Scholl, 2002; Jordan et al., 2003; Micheletti, 2003, pp. 92–93). One of the features that caught the attention of many intrigued observers was its inclusive governance arrangement, enabling the involvement of a wide range of stakeholders in the setting of labelling criteria (Scholl, 2002; Jordan et al., 2004).[11] As is the case with all labelling schemes, the Blue Angel faces clear limitations regarding the potential for market growth; but in comparative terms, it is well known among consumers and producers in Germany and abroad. The Blue Angel is used in public procurement strategies, is generally seen as an indicator of environmentally sound products and services, and has indirectly stimulated broader discussions of the environmental qualities of products (Scholl, 2002).

In 1989, The Nordic Council of Ministers established another well-known eco-labelling scheme: the Nordic Swan. They modelled it on the Blue Angel. It was the world's first multinational eco-label and is well recognized in the Nordic countries and abroad (Stø, 2002; Micheletti, 2003).

The European Union's eco-label (European eco-flower) is considerably less well-known than the Blue Angel and the Nordic Swan are in their respective countries. In 1988 the European Commission took an initiative to create an eco-label, but some time would elapse before it could be established. The success of existing labels, for instance the Nordic Swan and the Blue Angel, contributed to the difficulties for the growth of the EU eco-label (Jordan et al., 2004). The EU eco-label has sought to imitate the inclusive character of these schemes but continues to incur criticism from EMOs for being institutionally cumbersome, non-transparent, circumvented by claims from industry groups, and difficult to restructure (ibid.; Rubik, 2002).

In the United States, the US Green Seal, founded in 1989, has a dominant position in several product sectors. Aside from its market among private consumers, its market extends to large institutional purchasers, including government agencies, universities, and architectural building industries.

Our definition of green labelling may also include a number of *green trademarks*, provided they are based on standardization of prescriptive criteria. Green trademarks are clearly market-based and consumer-oriented and they are based on symbolic differentiation, but they are special in the sense that certain companies issue them on their own. They could be seen as self-labelling. To be sure, eco-labellers often distinguish eco-labels from green trademarks and may downplay these as greenwashing tactics and 'self-made promises' by superficially 'green' companies. Yet, conceptually, we can see these as green labels. High-profile green trademarks may also use labelling criteria as models for their own criteria. In the United States, for instance, Whole Foods Market (the world's largest retailer of 'natural' and 'organic' foods) has its own certification scheme, and in Sweden there is Coop's Änglamark label ('Angel Land').

## Other eco-standards

*Ranking and rating* are commonly used instruments in consumer, environmental and other informational policymaking. These instruments are similar to labelling in that they rely on differentiation. Schemes for ranking and rating can include many prescriptive criteria used for the assessment. Normally, a specific company has to reach a threshold or a specific criterion (or set of criteria) in order to be, for example, top-rated or ranked number one. Yet, the ranking and rating are used only as *information* (addressed, e.g., to consumers); in other words, they are not exactly addressed to companies as a *standard*, that is, a kind of rule that a company has to follow in order to be eligible for a certificate or licence. However, the difference between ranking and rating and labelling is subtle.

*Environmental management systems* (EMS) have been popular in many parts of the world. The ISO 14000 is considered the most widely recognized global-level voluntary initiative on the part of industry (Clapp, 2005). Such systems can be seen as 'standards of procedures' (Boström, 2003a), because they stipulate only that an organization must adopt and follow certain routines in order to improve its practice. The ISO 14000 family stipulates, among other things, that an organization must follow the law, have a policy for environmental practice, and follow routines to map, supervise, and measure environmental effects. A certified company should set its own goals for environmental improvement and make sure it progresses continuously. In contrast to green labels, the standards do not prescribe thresholds or requirements that must be achieved in order to qualify for a certificate.

In the environmental field, it is easy to observe plenty of eco-standards that lack prescriptive requirements, such as various *Codes of Conduct*. Codes of conduct are usually voluntary agreements, defined in a written document, which state the missions, values and practices that should govern in the marketplace. IGOs, such as the UN (Global Compact), have issued numerous codes of conduct, but they generally lack prescriptive criteria (Garsten, 2008). Many codes are self-developed by business actors and are therefore often attacked for their problems with vagueness and limited transparency and openness. As in ranking and rating, the difference between green labelling and these types of eco-standards is subtle, as our green mutual fund case reveals (see Chapter 5).

Another kind of eco-standard is *standards for reports and declarations*. Standards for reporting and declarations are based on rules for selection and presentation of information. Such standards do not stipulate appropriate or inappropriate performance, but specify the detailed information that the certified company has to provide concerning the socially or environmentally relevant aspects of the production processes. One example is the Global Reporting Initiative (GRI). This was initially an industry initiative on sustainability reporting for TNCs, but other types of organizations such as the 'Sierra Club', an American EMO, and the American Federation of Labor were among the founding members (see also Adams & Zutshi, 2005; Dingwerth, 2005). The GRI has since developed into a fully independent organization, with broad stakeholder involvement, and is now a collaborating centre with the UN Environmental Programme. It also cooperates with the United Nations on the Global Compact (Adams & Zutshi, 2005). We thus see broad support for this kind of standard, but there is also an emerging debate on eco-labelling vs. environmental reporting (declarations). This we discuss in the book in relation to our paper case.

All these standards are voluntary, explicit, codified, and written. However, with the exception of green labels, they do not stipulate prescriptive requirements, and they do not necessarily require an explicit or implicit consumer demand (they are not necessarily market-based and consumer-oriented), and they do not rely on symbolic differentiation, with the exception of ranking and rating (although standards for reporting and declarations may enable benchmarking). Although many policy actors endorse the existence of a diversity of eco-standards, these differences are also the subject of – sometimes heated – subpolitical (Beck, 1992) struggles among various kinds of actors. We argue that it is important to understand the diversity and the differences in order to understand arguments and debates about eco-standards. Later in this book, for example in Chapter 6, we will illustrate such arguments and debates. At the same time, we should not exaggerate the antagonism between proponents of different standards, because such standards are also framed as complementary.

# 4
# The Consumers' Role: Trusting, Reflecting or Influencing?

In the introductory chapter, we emphasized the broad societal embracing of the consumer role as one in which citizens should express their environmental and social concerns. This chapter begins by giving a brief overview of the academic literature that investigates such political and ethical consumerism – its ideas, patterns, and challenges. After the overview, the chapter suggests three ideal-typical views of the actual and potential roles of consumers with regard to informational and communicative tools, such as green labels.

## Green political consumerism

Basic research issues of green political consumerism are *whether, by whom*, and *how* the market can function as a political arena rather than merely as a realm for maximizing individualistic interests.

### Whether...

Fear of what consumption might do to a political community and to the public good is as old as consumption itself. Anxieties have been expressed in a variety of ideological traditions – from conservatism to Marxism (Soper & Trentmann, 2007, p. 3).

Underlying this 'whether' issue is a more normative one, which goes back – at least – to Locke in the seventeenth century (1689/1997; see also Sørensen, 2004), and which is tied to the *whether* question: should consumerism and market agency be seen as an ideal form of political empowerment, since virtually everyone has a penny to 'vote' with? The eighteenth-century economist Adam Smith believed so, but not to the extent that consumers should take 'externalities', such as social welfare, into account. Instead, Smith, who has many followers in today's neoclassical economics, contended that if all actors on the market, consumers included, acted 'rationally', by following only their individual (egoistical) interests and making their decisions on solid product information, the interests and goals of buyers and sellers would ultimately converge with societal interests, elsewhere called the common good (Smith, 1776/1974: 134; cf. Zwick et al., 2007). Another market-liberal economist, Ludwig Edler von Mises, echoed this view by providing a democratic image to describe the power of the consumer:

> The consumers are the masters, to whose whims the entrepreneurs and capitalists must adjust their investments and methods of production. The market chooses the entrepreneurs and the capitalists and removes them as soon as they prove failures. The market is a democracy in which every penny gives a right to vote and where voting is repeated every day. (von Mises, 1944, p. 17)

Although this position is very positive about the idea of consumer power, it implies – at least in today's versions – a criticism of exercises of consumer power where interests beyond self-interest are taken into account in purchases and other consumer actions. And today, despite all the media noise in which conscious, green, ethical, and value-based consumerism is embraced, other strong voices are heard criticizing such extended consumer concerns. In the influential magazine *The Economist*, for instance, such criticism is a recurrent theme. Behind their critical analysis of ecological and social implications of organic food and fair-trade coffee, for example, lies a deeper view, namely that markets become distorted if consumers try to achieve social and environmental improvements by choosing eco- and fair-trade-labelled products.[12] The magazine perceives green and ethical consumerism as a threat to open market competition that would make poor people (and threatened environments) everywhere better off in the long run. *The Economist* concludes the article with the following statement: 'Conventional political activity may not be as enjoyable as shopping, but it is far more likely to make a difference' (*The Economist*, 7 December 2006, print edition).

In addition to the neoclassical criticism of green political consumerism, there is the massive historical opposition of consumer power in general – from the political left and right. This is interesting in the light of the current embracing of consumer power across the political spectrum. In the nineteenth century, conservative groups were worried that society would be governed increasingly by 'Economic Man', 'with no intuition of unseen realities, no sensitivity to art or nature, no humility, and no inbred morals or sanction for their dictates' (Europe, history of, 2008). And on the left, Karl Marx divided humankind into (1) the political citizen (concerned about wider societal implications) and (2) the economic man (merely motivated by the individualist, Smithian rationality). It is intriguing to note how similar *The Economist's* neoclassical view is to Marx's radical scepticism concerning the market realm as the optimal one for making decisions with far-reaching sociopolitical consequences (cf. Klintman, 2002b).

While shooting at green political consumerism from the left, from the right, and from behind – if that is where neoclassical economics is located – one should not forget the criticism that comes from certain environmentalist voices. Although many environmentalists and EMOs talk favourably about green consumerism, others claim that such consumerism takes the wind out of the sails of more progressive environmental and social struggles, for instance legalistic endeavours and the activism of social movements. The ecological economist, Lintott, for instance, maintains that green consumerism, even if it has an environmental orientation and is 'modified', is still illegitimate and detrimental to the goal of a sustainable improvement of welfare, since a modified consumerism does not lead to *reduced* levels of consumption (Lintott, 1998, p. 240). These background issues are likely to be disputed for many years to come, with the use of theoretical as well as empirical developments in the research field, something that we perceive as a healthy part of a broader democratic debate in society.

Nonetheless, there are a large number of researchers and practitioners emphasizing the potential of green political consumerism. Like the opponents, they mix normative positions and descriptions, and it is beyond our scope here to try to separate the normative from the descriptive. Several major works on green and political consumerism are mentioned in Chapter 1 and in subsequent sections. Particularly interesting among the recent work, in our view, is the research that goes beyond the common observation that green political consumerism is increasingly being discussed, debated, and practised. While highlighting its positive potential of political consumerism, Micheletti and Føllesdal (2007) point out the risk that 'the exponential growth of voluntary codes of corporate conduct and labelling schemes' creates contradictions, incoherence in efforts, and superficial changes (Micheletti & Føllesdal, 2007, p. 167). Problematizing green political consumerism, without dogmatically rejecting it at the outset, is in our view the way forward for research at this point.

## Who...

Getting down to the more concrete *who* question of ethical and political consumerism, it is fair to say that the vast majority of studies have taken place – and still take place – on the 'front stage'. By front stage studies of ethical and political consumerism, we refer to studies, often social–psychological, of the motivation of various consumer groups concerning choices to purchase products and services marketed as advantageous in terms of reduced environmental, social, and animal-related harm.

We can see from previous studies that what researchers and market actors call 'concerned consumers' reflects quite heterogeneous consumer categories in terms of their motives and thoughts about alternative products. Much of the literature provides nice typologies of various consumers (Nordic Council of Ministers, 2001; Halkier, 2004; Worcester & Dawkins, 2005). The typical concerned consumer, according to several studies, is well-educated, belongs to the middle or upper-middle class, has a good income, is white, often a woman in her lower middle age or middle age, has children, and lives in Northern Europe or North America (see Goul Andersen & Tobiasen, 2004; Micheletti et al., 2004, 2005; Gilg et al., 2005; Strømsnes, 2005; Tobiasen, 2005; cf. Lindén & Carlsson-Kanyama, 2007).

Although researchers often conduct these studies in a solid, scientific manner, we argue that future research will have to go into depth and penetrate registered characteristics of various demographic consumer groups in order to arrive at a more nuanced picture of motives and consumer practices.[13] Nevertheless, a few results from these studies are quite relevant and important to raise for the purpose of this book. They relate to results about the demographic composition of political and ethical consumers – the *who* question – results that appear to be coherent across studies.

One key finding is that political consumers tend to be politically active in other arenas as well; thus these groups do not disregard other ways of performing politics (Micheletti & Stolle, 2005; Tobiasen, 2005). People categorized as political consumers are also more interested in politics in general. Accordingly, it is not mainly people who are very pessimistic and blasé about traditional political institutions who are engaged in political consumerism, as has sometimes been claimed (cf Harrison et al., 2005). Instead, the relationship is cumulative. People with an interest in politics are also interested in politics in the market arena.

It is also interesting to elaborate on the positive correlation that several studies indicate between high formal education and political consumerism (e.g., Micheletti & Stolle, 2005; Tobiasen, 2005; De Pelsmacker & W. Janssens, 2007). A reason for this positive relationship may be that well-educated people are trained and stimulated to follow the flow of information, codes, and symbols disseminated by media and experts, and communicated in official debates. The more we live in a 'knowledge society', 'information society', or even 'post-industrial society' (e.g., Nowotny et al., 2001), the more crucial it becomes for individuals to be able to handle – and select among – massive quantities of information, codes, and signs.

Nevertheless, we would hypothesize that well-educated people can also be the most ambivalent ones. Qualitative research on green consumption and everyday politics reveals some of the uncertainties and ambivalence of individuals' reasoning (Halkier, 2001, 2004; Sörbom, 2002). Bente Halkier maintains that '[a]mbivalence is the pervasive feature of consumers' constructions of their own roles as risk-handlers' (Halkier, 2004, p. 240). Furthermore, it is among 'green consumers' that one is most likely to find people with particularly reluctant attitudes towards green advertising (Zinkhan & Carlson, 1995; see also Crane, 2000).

There are two sides of the coin here (cf. Chapter 2 on individualization). On the one hand, the capacity of individuals has been strengthened historically by detraditionalization, secularization, an increased level of education, technological development, globalization, increased gender equality, and political modernization (Giddens, 1990, 1991; Beck, 1992; Nowotny et al., 2001). People become more reflective and learn to think critically about their own practices, and to question authorities. On the other hand, as society becomes more complex, risky, and differentiated, people become more dependent on expert systems (Giddens, 1990, 1991). People need more and more advice about how to live their lives in order to make better choices, but face dilemmas as they get contradictory advice from different experts (Höijer et al., 2005). Contradictory advice in turn leads to a demand for – and thus supply of – further advice; and so the spiral goes on (Bauman, 1991).

In sum, existing social theory and quantitative and qualitative research on political consumerism help us in developing a model of the typical political consumer, who is:

*reflective and self-reflexive:* he or she is well-educated, interested in politics, sceptical – but not disdainful – towards authorities. The typical political consumer is trained and socialized to revise his or her own previous thoughts and choices in the light of new information, although the political consumerist's behaviour involves a certain degree of routine.

*ambivalent and uncertain:* he or she has the capacity to choose on various bases, is educated in abstract thinking, drilled to make rational choices, and receptive to new information; but cannot know – at least not in the fundamental and absolute sense – if his or her choices really are a step ahead on the right track. He or she begins to question whether there really is such a single right track as policy actors now and then try to convince him or her.

*capable of developing reflective trust:* he or she may consciously choose whom to trust; he or she admits that some authorities have to do the standardization, although these can very well be non-state authorities. The typical political consumer seldom trusts anyone without reservation; it is not blind trust, although public bodies and well-known voluntary organizations are preferable to corporations. Yet, he or she does not have the time, motivation, energy or knowledge to scrutinize incessantly the state and non-state authorities that he or she has chosen to trust. Nonetheless, he or she may revise his or her trust because of new information, and is aware of that possibility.

**How ...**

There are obviously overlaps between studies that map out *who* practises green political consumerism and studies investigating *how* such consumerism is practised. Thus, we recommend the reader to use the above-mentioned references to obtain some answers to the *how* question. This section gives a few additional comments. Political green consumerism is often exemplified by the practices of boycotting and buycotting.

Boycotts (see Friedman, 1999) are often seen as one of the original practices (or non-practices), with examples from ancient times to today. In terms of theory development, boycott studies are useful in that they help to remove a political halo over political consumerism as one entity. Examples of Nazi boycotts of Jewish goods, along with other boycotts based on racism, elucidate how political consumerism is certainly value-based, although not necessarily ethical in a democratic, humanist sense. Furthermore, what is interesting is that research on boycotts has so far primarily referred to avoidance of products produced by specific companies, countries, or groups. Researchers usually miss (whether intentionally or not) what is in our view consumers' more radical and thoroughgoing avoidance of, for instance, petrol, meat from frame-raised cattle, or nuclear power.[14] The term boycott does not usually capture avoidance of entire product categories. It is often used too narrowly, denoting a number of specific, often time-limited avoidance practices (albeit with a potentially immense impact on the boycotted companies, countries, or groups). Boycotts of sweatshops are an important exception, which research illustrates as long-term and broad-ranging, with an impact on a whole type of industry (O'Rourke, 2005).

*Buycotts* are the sister practices of boycotts, referring to active consumer choices of products and services which consumers perceive as being in line with their values. Research information about buycotts is frequently collected and reported in the same articles and reports where boycotts are examined. Consumers' choices of products with green labels are one type of buycotts, along with products and services with other types of information that certain consumer groups are particularly positive about.

Both boycotts and buycotts are *monetary* types of green and political consumerism (see Table 4.1 below). Still, green and political consumerism is not confined to consumers' concrete monetary transaction to acquire a product or service or to the green and political considerations that precede the transaction. To capture the essence of the phenomenon, we have to acknowledge the broader aspects of political and ethical consumerism, namely *the discursive and communicative action* performed by individuals and organizations in order to alter and develop products, production processes, and consumer-related policies based on political and ethical concerns (Micheletti & Stolle, 2005). Consumer protests in word and writing, consumer demonstrations, and Internet activities

*Table 4.1* Types of green political consumerism

| Green political consumerism | Monetary | Discursive |
| --- | --- | --- |
| Frontstage | Consumers' boycotts or buycotts of products and services based on the green and political values of these consumers | Protests, demonstrations, and communication in media and Internet about products and services based on the green and political values of these consumers |
| Backstage | Consumers' monetary support to people and organizations that promote green political consumer issues | Consumers' participation, involvement, and debates in IGOs, NGOs, SMOs, and in public debates about political, consumer-oriented policy tools and practices |

through debates and campaigns to influence companies and other consumers are part of what can be called discursive political consumerism. Discussions about the discursive side of political consumerism are parts of the much broader public and scholarly debates about the preconditions for new modes of governance, and for inclusions of more deliberative elements in a strengthened democracy. In the *who* section of this chapter, we mentioned the common focus on the front stage in research. This is true concerning monetary, political consumerism. However, it is also mainly the front stage that is in focus in studies of discursive consumerism. Consumer protests, demonstrations, Internet reactions, and so forth, are usually media-oriented events where final product types as well as particular companies and countries are in focus. Less research has been devoted to discursive and communicative practices on the back stage, where the actors involved develop, dispute, and organize consumer policies, and where marketing strategies for such tools are chosen. In the following sections, we introduce a model of the relation between policy tools and consumer roles, a model that seeks to acknowledge both the front and back stages of green political consumerism.

## Three views of policy tools and consumer roles

Here, we suggest distinguishing analytically between three different views of green consumer empowerment and roles. We have used findings from our case studies on green labelling when we developed these, but we would like to emphasize that it is more adequate to consider them as theoretical constructs. In subsequent chapters in the book we will compare these ideal-typical constructs with empirical examples and normative discussion (see, e.g., the concluding chapter). Our comparison of these views with the literature on participation, deliberative democracy, and consumer empowerment has been the basis for three ways of conceiving the roles that political consumer tools should play (see table 4.2): the ideals of *simple trust, insight,* and *influence* of consumers through these tools. Simple trust refers to an emphasis on consumers' need for unambiguous and pragmatic information. *Insight* stands for giving priority to detailed information, and for the importance of disclosing uncertainties to consumers. *Double influence*, finally, implies the view that successful green consumer empowerment – in addition to the influence through selective shopping – needs forums where consumers can react and protest as a collective to influence other stakeholders. Importantly, these three ideal types should not be

*Table 4.2* Views of labels and consumer roles

| Views of labels and green political consumers | Simple consumer trust | Consumer insight | Consumer influence |
|---|---|---|---|
| Goals | consumers who trust green and political consumer policies | enlightened and reflective political consumers | green and political participatory consumers |
| Types of Instruments | simple labels and standards | several systems of labelling | participatory labelling systems |
| Idea Behind Instruments | pragmatic: motivate broad consumer groups by avoiding confusion | disclose uncertainties and diverging views, reduce consumer naïvety | stimulate consumers' impact, avoid captured, passive consumers |
| The Value of Knowledge | basic knowledge about stated aims of labels | substantive and procedural transparency is needed | knowledge needed as a basis for participation and debate |

confused with positions or points of view of specific groups or persons. Few actors would fully subscribe to one of these three. Rather, different actors are likely to put different relative weight on the three ideal types. Nor should it necessarily be assumed that these represent a normative scale, for instance from 'unsatisfactory' to 'very satisfactory', in our view. Instead, we see the three ideal types as bases for our further normative exploration.

## Political consumerism based on simple trust

In many of the calls for consumer trust and comfort in tools of political consumerism, the stated challenge is to create labelling and information that stress the stability of the label. To create stable standards at national or international levels is usually regarded as the best way to establish consumer trust and avoid confusion (Elkington et al., 1990; Organic Trade Association, 2002). Thus, the complexities of the social and political procedures that underlie labelling criteria and standards are not very relevant for consumers to learn about. These procedures, along with the substantive complexities of the actual decisions on criteria, would run the risk of creating consumer confusion and a sense of arbitrariness along with an excessively relativist view of labelling, that is, that no labelling scheme or claim is better than any other (Fernau, 2001). In this view, consumers need a basic awareness of what the label stands for – the 'eco-friendliness' of a product, for instance. This basic awareness ought to motivate broad consumer groups to choose the labelled products. If this is the case, the label is considered a successful marketing tool aimed at, for instance, increased international trade, and perhaps a more progressive market which takes green and societal issues into account (Rhodes & Brown, 1997). Here we see the pragmatic view that creating trust among a larger number of consumers has several advantages (economic, environmental, improved health, etc.), even if the labelling criteria and procedures are not fully understood or even if they are more compromised than some consumers might appreciate.

If the simple trust dimension of political consumerism succeeds without the other dimensions, it may lead to a certain, albeit low, procedural and substantive awareness among many consumers. Broad consumer groups may become 'political' through their choices of goods, but few are likely to go further in their engagement as political actors. In political terms, this dimension would be a direct democracy only in the sense of consumers becoming 'voters' on a single issue, for instance, for or against 'green electricity'. However, it would not be democratic in the sense that the consumer would be aware of the specific criteria of 'soundness' or of the decision-making procedures behind the labelling criteria. It would thus be a direct, uninformed, and non-participatory, activity.

### Insight-based political consumerism

When emphasizing the importance of consumers' insight into the political consumerist tools, the key challenge is rather how to create labelling and information which describe and analyse complexities. Information about the tools should include the substantive context. The label itself is a basic separator of products and processes, which needs a comprehensive complement in information and open debates (e.g., Erskine & Collins, 1997). Transparency is a buzzword here. It may be the case that the goal of one national or international standardized label is too crude for consumers who have become more individualized through insight into the products and processes, or at least have been divided into political consumer groups (Allen & Kovach, 2000, pp. 221, 228).

Perhaps for this reason there is a need for several competing labelling systems at several levels, similar to political parties (cf. Karl & Orwat, 1999). In any case, there is a need for open and continuous revision of the label criteria through open debate. In addition to the substantive context, the insight-based view of political consumerism may include the importance of consumer awareness of the political and strategic *procedures* that lead to certain labelling criteria (Pepper, 1996). The argumentative framings, rhetoric, and flexibility of preferences are crucial factors here.

The consumer confusion that many actors fear – particularly actors who endorse the simple trust approach – is regarded by insight-based consumerist proponents as a necessary reflection of a multifaceted reality. Those in favour of insight-based consumerism hold that confusion may ideally stimulate political consumers to learn more (cf. Eder, 1996, p. 154ff; Yearley, 1996, p. ix; cf. Klintman, 2002a). Such increased learning , in turn, may generate consumers who – through their active choices of goods – use or relate to the labels and standards with insight into the uncertainties and complexities behind these instruments. By extension, such insights may have advantages for the democratic, social, and environmental implications of consumerism; a more informed political consumer acts democratically in a deeper sense. A more informed political consumer is also likely to consider external aspects – health, environment, social justice, animal welfare – thus enhancing the extrinsic value of consumerism (Wasik, 1996, p. 93ff.).

Compared with trust-based consumerism, insight-based political consumerism may ideally lead to a high procedural and substantive awareness. Still, one should note that consumer insight does not need to entail a profound knowledge of *the content* of ecological and ideological complexities. To see *patterns* in the way the field is complex, partly uncertain, and ideological is a good step forward in consumer insight. However, it is arguably less likely that such insight may reach as many consumers as may the consumerism based on simple trust (cf. Nimon & Beghin, 1999). Perhaps broad consumer groups may become 'political' through their choices between labelled and unlabelled products, whereas only a narrower fraction of consumers are likely to gain the insight called for in this view of political consumerism. In political terms, this insight-based type of consumerism is a direct democracy in the sense of consumers becoming 'voters' on a single issue, for instance, for or against SRI funds. Moreover, this type would also be democratic in the sense that the consumer would be aware of the specific issues defined as 'sound', or of the substantive reasoning behind the labelling criteria. It would thus be a direct, informed, albeit non-participatory, democracy.

## Political consumerism based on double influence

Aside from the influence that consumers exert by purchasing goods based on labels and standards, the view of political consumerism that stresses 'the double influence' implies an endorsement of consumer participation in a broader sense, in addition to the endorsement of politically selective shopping. Accordingly, an important political consumerism ought to include consumers' active involvement in the substantive, and perhaps procedural, policy contexts of consumerist tools (read: discursive political consumerism on the front and back stages). Thus, political consumers should be able to have a strong influence on labelling criteria, priorities, and concerns that constitute consumerist tools. Moreover, consumers should ideally be able to influence the political procedures behind the labels – reduce obscurity, require open deliberation, and so forth, behind the labels (Klintman, 2002b; cf. Zavestoski et al., 2006). In addition to the participation of NGOs, there should be forums for broad public involvement in the substantive factors of consumerist tools. If the influence-based consumer ideal is separated from the other two ideals, its advantages can mainly be argued by emphasizing the intrinsic value of consumer engagement. The fact that consumers make political choices – not only through their selection of goods but also through their broader involvement – is highly appreciated by those who equate more consumer decisions with more democracy.

Compared with insight-based political consumerism, influence-based political consumerism supports a high procedural and substantive influence. However, like the former, influence-based consumerism is less likely than political consumerism based on simple trust to involve but a smaller fraction of consumers (cf. Dryzek's (2001) examination of the tension between deliberation and representativeness). In other areas (consumer areas and elsewhere) active political protests have rarely involved a majority. Thus, there is a risk that the few loud consumer protesters are conceived as representing the voices of most political consumers.

In political terms, this influence-based type of consumerism is direct democracy in the sense of consumers becoming 'voters' on a single issue – for instance, for or against organic food processes. Moreover, this type would be democratic in the sense that the consumer may exert an influence on individual, conspicuous, aspects of the consumerist tools.

Yet, there is a risk of a narrow and crude awareness of one factor. This type of consumerism does not necessarily have to prioritize a broader insight into the substantive and procedural context. It may be based on the influence of charismatic authorities representing a particular view. Procedural insight into strategies behind campaigns and so forth would in such a case be even less likely. The result would therefore be a direct, possibly uninformed, participatory democracy. On the other hand, one should mention that some social thinkers, for instance John Stuart Mill, claim that public participation always has a certain beneficial educative effect on citizens.

## Conclusion

This chapter has provided an overview of ideas and research about green political consumerism. For elaborating on how some of the studied labelling processes can be judged in terms of political consumerism, the three dimensions of consumer roles and empowerment, we argue, offer a useful point of departure. Such a judgement of labelling schemes depends on how the actor with the ambition of making such an evaluation views the relative importance of simple trust, insight, and influence. The three views are theoretical constructs. We have described them in a way that makes it difficult to be fully content with any one of these. Simple consumer trust in green policies may appeal to many consumers, but runs the risk of consumers being victims of greenwashing, or showing categorical mistrust in green consumer policies. Insight-based political consumerism may lead to highly enlightened consumers, although most likely only to marginal consumer groups, who remain more academic than active when it comes to improvements of green consumer policies. Political consumerism based on double influence is promising through the high amount of political consumer activity. Yet, without an insight-orientation, these activities run the risk of being rather simple yes-no reactions, without well-reflected suggestions of constructive policy alternatives. We aim to relate these consumer roles to findings in subsequent chapters. Moreover, we use these consumer roles in the concluding chapter, where we discuss reflective trust and consumer empowerment in practical policymaking.

# 5
# Our Cases

This chapter gives an overview of our cases. We describe the history behind the various schemes and their current position, and we discuss how we have selected our cases. The chapter gives the reader an empirical basis that is useful when reading subsequent chapters, where we analyse the cases more thoroughly.

## Organic food labelling[15]

An early pioneer in many countries in alternative agriculture production was the biodynamic movement, developed from Rudolf Steiner's anthroposophy. The movement developed the first standards for Demeter quality control as early as 1928.[16] However, it was not until the 1970s that alternative agriculture in various streams enjoyed a noticeable upswing. For example, in the United States in that period, alternative agriculture rose from a small-scale business in the periphery to become at least big enough to be subject to regulation (Golan et al., 2000, p. 27). Ecological farmers were still seen as a small group of devoted idealists (Lathrop, 1991), but the alternative agriculture movement appeared as a palpable part of a growing radicalizing environmental movement. This movement provided a broad, thorough critique of industrialized societies and large-scale capitalist business. Schumacher's (1973) motto 'small is beautiful' fitted the ideals of alternative agriculture very well. Frames around small scale, local production, and self-sufficiency were used in protests against large-scale production, distribution, structural rationalization, and the chemical-based conventional agriculture. As part of this movement, the International Federation of Organic Agriculture Movements (IFOAM) was established in 1972 by five organizations: Nature et Progrès (France), Soil Association (UK and

South Africa), Rodale Press (USA), and the Biodynamic Association (Sweden).

Alternative agriculture, including the organization IFOAM, remained small in the 1970s. Then most Western countries gradually underwent a development in which organic movements initiated standards for organic production (Torjusen et al., 2004). IFOAM grew rapidly in the 1980s with many new member organizations. It set its Basic Standards for organic production and assisted in the development of national standards. It further developed contacts and lobby units in relation to such international organizations as the EU and the FAO. IFOAM had a strong role in the work towards an EU regulation on organic farming that was introduced in the early 1990s (Le Guillou & Scharpé, 2000). Furthermore, national private labelling organizations adopted standards, or they were incorporated in state rules. In many parts of the (rich) world, organic farming and organic labelled foods are becoming big business. It is one of the fastest-growing segments in the consumer goods markets in many countries (Micheletti, 2003, p. 98).

### Organic labelling in Sweden

In Sweden, four organizations within the Swedish organic movement established KRAV in 1985 (KRAV, 2000). KRAV started to issue rules for organic production and to monitor licence holders. It was accredited as a controlling body according to the IFOAM standards. The Swedish Ecological Farmers largely controlled KRAV's activities during its first five years, almost completely through voluntary work. Although several key players in the field – including authorities, conventional farmers, and processing industries – ignored or were suspicious of KRAV in its early years, the organization gradually received a more positive response. This was partly due to a proactive role played by large Swedish food retailers (KF and ICA) that supported KRAV early on. The retailers quickly realized the importance of an independent and credible third party that could scrutinize the environmental claims of products and labels, given the recent rise of many self-made 'green' symbols and labels on product packages.

In the 1990s, KRAV grew rapidly and incorporated many interest groups into the organization. By 2007, KRAV had 28 member organizations reflecting different interest groups, for example, the food industry, environmental and animal protection organizations, food retailers, unions and farmers' associations – both the main association for conventional farmers and the Swedish Ecological Farmers. KRAV's members are represented on the board and in committees. Accordingly,

KRAV is a hybrid organization that has managed to include most big Swedish players in the field of agriculture, food production, and food distribution. Although state authorities are not included as formal members, KRAV has gradually developed a closer relationship with the state. KRAV has state authorization to audit organic agriculture and production and to ensure that EU regulations about organic production are followed. At the same time, KRAV provided and continues to provide its own stricter standards, covering a broader spectrum of issues than those of the EU regulation. However, the centralization of the EU regulation has recently challenged the autonomy of this and similar types of organizations across Europe (see Chapter 7).

In general, there is high citizen awareness of KRAV, and the Swedish public associates the KRAV label with products that are good for health, the environment, and animal welfare (Magnusson et al., 2001; Ekelund, 2003). In the mid-1990s, KRAV expanded quite dramatically and the trend remains positive in the 2000s. In 2006, 6.6 per cent of all Swedish arable land was KRAV-certified and about 3 per cent of the Swedish total food consumption (sales value) was organic food.[17] KRAV's criteria have been developed and expanded continuously, covering new types of food products (e.g., processed food, seafood). Although this is often, and to some extent correctly, seen as a 'success story', we will later in the book discuss labelling challenges such as rigid framings (Chapter 8) by referring to this case.

### Organic food labelling in the United States

The history and characteristics of organic labelling in the United States are quite different. It has therefore been interesting, for comparative reasons, to select this as our second case. In 1973, Oregon was the first state to pass a state law regulating organic food as a response to reports on fraud and inconsistencies regarding organic claims. Several other states soon followed, but substantial differences in state organic farming regulation arose across the United States and certain states did not require third-party certification. Instead, it was up to each company to label its own products as 'organic'. Moreover, according to Amaditz (1997), the most serious problem before 1990 was that producers and marketers in several (28) unregulated states could continue to make claims that were arguably capricious when compared with the other states' definitions.

The broad picture of American organic food labelling in the 1990s and 2000s is that of a federal ambition to move from a complex and diverse system to a nationally standardized one. This ambition was manifested by the incorporation of the Organic Food Protection Act (OFPA) into the 1990 Farm Bill. Some of the explicit goals set by Congress were, as in the EU, to assure consumers that organically produced products meet a consistent standard and to facilitate interstate commerce in fresh and processed food that is organically produced (Klintman & Boström, 2004; Boström & Klintman, 2006a). According to Amaditz (1997), there was a consensus across a broad range of actors (e.g., state agriculture departments, national farmers' organizations, organic industry, trade associations, and consumer interests) that the United States needed a national standard for organic food. A strong incentive for the standardization was also the increasing interest among American agricultural actors in moving closer to EU organic standards for economic and trade reasons (Golan et al., 2000). The trade motive can be linked with the steady growth rates of US organic food sales of 17 to 21 per cent between 1997 and 2003. The sales in 2003 represented 1.9 per cent of total US food sales (OTA, 2004). Although this is still a fairly modest part of total food sales, the steady growth makes organic food sales an important part of food trade in the United States.

The US Department of Agriculture (USDA) authorized a National Organic Programme (NOP) to develop a 'National List' of Allowed Synthetic and Prohibited Non-Synthetic Substances for organic production, labelling requirements, an accreditation programme, and guidelines for imports and exports (Frankel & Borque, 1998, p. 1; Alternative Farming Systems Information Center, 2001). Despite increasing EU influence on Swedish organic labelling, the US situation reveals a more (federal) state-led and centralized standard setting in this sector. We return to this difference in subsequent chapters.

What is central to the understanding of the debates in the United States and our Swedish–US comparisons is also a board called the National Organic Standards Board (NOSB). The board has the task of developing standards (within NOP) and giving recommendations to the USDA. The NOSB has individual members from different organic-food-related backgrounds, who should all be organic experts and environmental and consumer advocates. Whereas KRAV has members from around 30 *organizations*, the number of members in NOSB is more limited (approximately 13–16 *individuals*). In addition, whereas the KRAV members include those who are only partially involved in organic practices (and partly in 'conventional food practices'), the NOSB encompasses members who are mainly associated with organics. A third fundamental difference is that KRAV (although an NGO) has the authority to set its own labelling standards as long as it does not set lower standards than those regulated by the EU. In contrast, NOSB just gives recommendations to the USDA, which may accept or reject the recommendations. Such differences create interesting comparative opportunities, which we shall have reason to return to.

## Forest certification and labelling[18]

Forest certification and labelling represents one of the most intriguing examples of green labelling worldwide, and the establishment of the Forest Stewardship Council (FSC) in 1993 has received extraordinary scholarly attention for its novel way of organizing green governance (Domask, 2003; McNichol, 2003; Gulbrandsen, 2008). It is highly inter- esting for our purposes, partly because of Sweden's front-running role in this case and the contrasting US case.

The historical threads of forest certification and labelling are shorter than those of organic labelling. An emerging global forest crisis since the late 1970s preceded market-based initiatives in forest policy in the 1990s (e.g., Elliot, 1999; Bendell & Murphy, 2000; Domask, 2003; Dingwerth, 2005). Such threats as tropical deforestation, the loss of old- growth forests in temperate and boreal zones, and biodiversity loss fuelled social movements in their targeting of forest-related industries. Forest issues received considerable media and public attention. Indigenous civil society groups in the South and EMOs such as Greenpeace and FoE in the North staged protests. EMOs targeted gov- ernments and retailers in the Do-It-Yourself (DIY) and furniture mar- kets. 'Customers began to write letters to the retailers and to confront store managers and employees with tough questions about timber sourcing' (Bendell & Murphy, 2000, p. 69). Boycott campaigning led certain big image-conscious retailers to stop using tropical forest prod- ucts (ibid.; McNichol, 2003). Many municipal governments in the United Kingdom, the United States, Germany, and the Netherlands banned the use of tropical forest products in their procurements (Domask, 2003). As a consequence, for many DIY retailers and suppliers, EMO-supported certification and labelling appeared to provide a solution to an escalating business problem.

Disappointment with inert intergovernmental regulatory processes gave additional incentives to the search for market-based solutions. IGOs failed to make a visible impact on global or regional deforestation rates, so the WWF, the World Conservation Union (IUCN), and the UNEP recommended certification in the second edition of their 'World Conservation Strategy' published before UNCED in 1991. Failure among

UN states to reach a binding forest convention in Rio the year after added to this critical view.

FSC was established as an international organization in 1993 for the purpose of promoting sustainable forestry around the world. The FSC defined ten principles that should guide responsible and sustainable forestry in any local context.[19] The members of FSC represent *social interests* such as labour unions and groups representing indigenous people, *environmental interests*, and *business interests*; and, formally, each of these three interest groups has an equal, one-third, share of the decision-making power. In addition, the voting power is divided equally between developed (Northern) and developing (Southern) country members in each of the three chambers. Regional and national standard-setting processes – in which regional and national specific interpretations of the FSC's general principles and criteria are developed and implemented – must follow this basic organizational structure. By 2007 FSC had forest management and chain-of-custody certificates in 84 countries.[20]

### FSC in Sweden

Sweden was the first country to introduce a nationally adjusted FSC standard. Its introduction was the result of intensive campaigning by parts of the Swedish environmental movement: WWF Sweden and the Swedish Society for Nature Conservation (SSNC). Their initiative met with criticism and counter-moves from forest industries and owners, but through the building of a pro-FSC network that included purchasers of forest resource products (e.g., IKEA), the EMOs managed to establish an 'FSC working group' in February 1996, which, among other interest groups, included all the large forest companies in Sweden (see Elliot, 1999). Political organizations and state agencies were not allowed to participate as members of the working group, but the relationship between the FSC network and state authorities was friendly and collaborative, which facilitated the implementation (Boström, 2002, 2003b).

Interviewees from both EMOs and the Swedish forest industry also declared how significant it was for broad participation that Swedish EMOs addressed the severe situation in the Swedish forests at the Rio Summit and other international meetings in the early 1990s. This global publicity was important since the large Swedish forest companies export a major part of their forest products, both timber and paper products. Swedish EMOs cooperated with EMOs in England, Germany and the Netherlands, which in turn campaigned domestically and organized groups of buyers, including key retailers. These retailers demanded verification by Swedish forest producers that the forestry was going to be managed in a sustainable way.

After a long process of negotiations and compromises, the groups agreed on a standard proposal, and the FSC's international board approved it on 5 May 1998. Two of the participants, Greenpeace and Skogsägarna (the association for private forest landowners), withdrew from the process because they could not reach a compromise over some of the criteria. Their withdrawal was disappointing to many involved, but the Swedish situation was nonetheless unique, due to the major agreement among a large part of the environmental movement, social interests and a large share of business interests.

Nearly half of the Swedish productive forest land is certified in accordance with the FSC standard, which in the early 2000s constituted about 30 per cent of the world's FSC-certified forests (Boström, 2003b). In November 2007, it is about 12 per cent, which is due to rapid increase in many other countries.[21] In many other parts of the world, EMOs have not been that successful in persuading the forestry industry about the benefits of FSC certification. In several nations and regions, a competing industry and/or landowner-dominated standards – such as the Programme for the Endorsement of Forest Certification schemes (PEFC) – have marginalized the FSC (Ozinga, 2001; Cashore et al., 2004; Gulbrandsen, 2004). Opponents developed the competing PEFC in Sweden, with non-industrial forest owners among its chief members, but this counter-move has not threatened FSC's dominance so far.

### The United States – FSC and the competing Sustainable Forestry Initiative

If the Swedish case could be seen as the 'success story' of FSC implementation, the American case could be seen as a 'success story' of an FSC competitor, the Sustainable Forestry Initiative (SFI). In the United States, SFI forcefully marginalized the FSC, despite the fact that many of the groups that participated in the establishment of the global FSC actually came from the United States (Synnott, 2005). In addition, several initial meetings preceding the establishment of the FSC were held in the United States. Nevertheless, the FSC did not appeal to the US forest sector. The forest companies were particularly concerned about the FSC's wide-ranging performance-based approach to forest management, the stringency of the standards, its chain-of-custody requirements, and the variations in the region-specific FSC standards[22] (Cashore et al., 2004).

As a response to the FSC initiative, US forest industries quickly mobilized support for a competing programme. Instead of rejecting certification altogether – a move that many US forest companies would indeed have supported – a proactive response to the idea of certification was promoted. Their recently established trade organization, the American Forest and Paper Association (AF&PA), proved to be an effective organizational channel to promote the SFI forcefully. EMOs and other FSC supporters immediately saw the SFI as detrimental to all ambitions towards sustainable forestry by way of stringent certification standards. They criticized the programme for lacking performance standards, transparency, a chain-of-custody system, and mandatory third-party audits. The SFI was framed as the logging industry's attempt to 'self-regulate' (Cashore et al., 2004).

The AF&PA responded in an aggressive manner and designed public campaigns to address its dissatisfaction with the FSC. The trade association successfully mobilized support from public bodies and states. Several state legislatures enacted resolutions of proclamations endorsing the AF&PA and/or the SFI programme. The SFI also received support from many retailers. Moreover, the AF&PA strategically focused on arenas in which the FSC seemed to be making limited progress. It also directly lobbied companies with pro-FSC procurement policies to change their policy wording to allow the acceptance of SFI-certified products.

All the AF&PA members are required to adhere to the SFI Principles and Implementation Guidelines, and since 1998 non-AF&PA members can also use the SFI certification. The SFI standard has been developed gradually, as we analyse in Chapter 9, and SFI eventually became a member of the PEFC to signify its international orientation. According to Benjamin Cashore and colleagues (2004), who investigated the introduction of forest certification in the United States, Sweden, Germany, the United Kingdom, and British Columbia in Canada, the divisions between FSC and SFI in the United States created the most polarized climate in all their cases. Hence, the Swedish FSC case and American SFI counter-move give important input to our analysis of the consensual versus adversarial political cultures in the respective countries.

## GMs: no mandatory labelling in the United States, but labelling in the EU[23]

In this book we mainly focus on voluntary labels. However, it is important to note that some labels are mandatory, such as requiring information on ingredients and country of origin on product packages. These examples are not green labelling in the sense we employ here. However, the EU has established several mandatory labelling regulations, which have interesting contrasting features. One example is mandatory GM labelling in Europe. By comparison, one may note that the United States does not require labelling systems that inform about GM processes behind final food products. Yet, there has been intensive debate regarding whether such a mandatory label should be introduced in the United States as well. In our research projects, we have followed mainly the US case particularly because we expected the US labelling debate would give us important input to the arguments for or against labelling as such.

In the United States, where there is no mandatory GM labelling, it is the principle of 'substantial equivalence' rather than the precautionary principle that dominates. If two products or production processes are not proven 'substantially unequivalent' it would, in this view, be unfair, in terms of competition, if one of the products had to use a label that consumers might perceive as a warning. Interesting challenges emerge when consumers and NGOs express concerns that go beyond this 'yes, unless' principle. In the policy issue of mandatory GM labelling in the United States, consumers and NGOs have indeed expressed concerns. One concern is the risk uncertainty involved. Who can know for certain that no negative consequences will emerge from this new technology in a couple of decades, for human health or for various parts of the environment? Aside from the scientific disputes about risk uncertainties, it is intriguing to note that the US government and the GM coalition base their arguments against mandatory GM labelling on the notion that the only concerns relevant to consumers are those of classical Economic Man: price, nutrition, and the general quality of the final product. All extra information is economically irrational, such as information about food processes. Moreover, the GM coalition also maintains that such labelling is severely *misleading.* Thus a mandatory GM label would not increase consumer choice; it would reduce consumers' freedom of choice by its irrelevance (see the analysis of the controversy in Klintman 2002a, b).

In the EU, the regulatory development of GM appears to be entirely different from that in the United States. Already in 1978, the Commission proposed through Directorate-General (DG) XII (Science, Research, and Development) that notification and authorization be required by national authorities prior to all work and research that involved recombinant DNA (rDNA) (Rosendal, 2005). After further directives on the development of deliberative release of GMs in the 1980s and 1990s, a new directive sharpened regulations further, partly by amendments to mandatory labelling (Directive 2001/18/EC). Similar to regulations on food additives and flavourings, there are at least two principles for GM foods in the EU: they should be safe for consumers and the environment, and they should allow the end consumer to choose whether or not to eat GM foods (Cheftel, 2005). This means that all products that consist of, or include, GMs should be labelled. In addition, food should be labelled if it is produced using GMs but processed in such a way that the ultimate product does not contain GM-based DNA.

Like most other labelling systems, the GM label is not absolute; labelling is required only if the GM content exceeds 0.9 per cent, or if the inclusion of GMs has been unavoidable (2001/18/EC, articles 21 and 26.1).[24] They give a tolerance level of 0.5 per cent for accidental presence of GMs, below which labelling is not required (1830/2003 article 4, pp. 6–8). Nor do milk, eggs, and meat from animals fed on GM feeds need to be labelled.

Several controversies have taken place within the EU, and externally, not least through cross-Atlantic disputes. Internal disputes concern how to deal with the coexistence of organic farming, conventional farming and GM plants, in terms of traceability and burden of proof (Soneryd, 2008). Externally, a moratorium in the EU on marketing approvals of GMs has been a heated issue *vis-à-vis* the United States. Since 1998 a moratorium has blocked EU imports of several GM products, with much complaint from the US food industry (Rosendal, 2005). In the view of the EU, it has been beyond the responsibility of single consumers to decide whether they want to purchase these products, largely because experts in the EU perceive the underlying production processes as an environmental risk.

Several policy researchers have taken an interest in this difference in regulatory stringency concerning GMs in the United States and the EU (Levidow & Boschert, 2008). They have given several explanations. One refers to a lack of unity and of a common strategy within the GM industry. Another explanation is the strong and effective coalition between environmental groups and EU institutions in favour of strict regulation, which has paved the way for strict GM regulation (Patterson, 2000).[25]

## Marine certification and seafood labelling in Sweden[26]

In comparative terms, both the Swedish organic case and the Swedish FSC case are often seen as rather successful in terms of implementation and broad stakeholder involvement. We wanted to include a Swedish case that appeared more conflict-laden and difficult to carry through, and in that way be able to test some of our working hypotheses. For instance, could it be the case that we exaggerated the impact of the consensus tradition in Sweden by analysing only these previous 'success stories'? When we decided to devote research time and resources to this case, the seafood labelling debate appeared to be at a stalemate.

The main international organization associated with marine certification and seafood labelling is the Marine Stewardship Council (MSC). The backdrop to the MSC's establishment relates to the global crisis in fisheries and regulatory paralysis. A painful experience some years before the establishment of the MSC was the collapse of the Grand Banks cod fishery of Newfoundland, Canada, and the loss of over 40,000 related jobs in the industry (Howes, 2005).

As in the forest case, EMOs criticized IGOs for paralysis, regulatory failure and inability to tackle effectively the overcapacity in the global industrial fishing fleet. According to Michael Sutton of the WWF: 'This history of fisheries management is one of spectacular failure. By working together with progressive seafood companies, we can harness consumer power in support of conservation and make it easier for governments to act' (quoted in Constance & Bonanno, 2000, p. 129). There was an important buyer in the business that began to think the same. Unilever, which is one of the world's largest buyers of frozen fish, recognized that, 'unless major fisheries took stronger steps to become sustainable, the company would no longer have access to its valued raw material' (Weir, 2000, p. 119). WWF and Unilever formed the MSC in 1997 (Fowler & Heap, 2000), which has since developed into an independent, multi-stakeholder organization, and about two dozen fisheries around the world are now certified according to the MSC standard. Labelling is made possible through a chain-of-custody standard.

WWF Sweden had good, fresh experiences of working with the FSC case and thought it would be possible to introduce the MSC in Sweden also. However, the MSC was heavily criticized by fishermen and public authorities from Scandinavian and other countries (Constance & Bonanno, 2000; Boström, 2004b, 2006a). They thought the MSC was designed to replace or circumvent existing democratic institutions, thereby bypassing governments, and they were concerned primarily because a private transnational corporation (Unilever) was a main sponsor and played a central role. WWF ceased to exert pressure to introduce the MSC in Sweden because of all the negative attitudes.

The MSC nevertheless had an important impact in Scandinavia in that it spurred a process with various efforts to introduce labelling schemes for seafood (see also Nordic Council of Ministers, 2000). After several years' heated and unproductive meetings, a breakthrough happened as soon as KRAV (the Swedish organic labelling organization described above) offered to coordinate the development of standards. Two years of standard draftings and intensive stakeholder debates eventually led to a final standard proposal that was approved by KRAV's Board of Directors in March 2004. Since the autumn of 2004 a few fishing vessels have been certified for green labelling of shrimps and herrings; no strong market impact has been felt.

What is especially interesting about this case are the serious initial controversies and mutual mistrust despite the Swedish consensus culture. In subsequent chapters, we aim to analyse – and compare with other cases – how it was possible to reach agreements among groups despite these poor circumstances, and what kind of agreements it was possible to reach.

## Green electricity[27]

To understand green labelling of electricity, one ought to examine how it is intertwined with several other policy and business strategies aimed at developing green electricity policies. Other policies include subsidies/tax incentives, green certificate schemes, and renewable energy funds. Since the space in this book does not allow for such a comprehensive outlook, the reader is recommended supplementary reading that will help develop a full picture: del Río & Gual, 2004; Vachon & Menz, 2006; Gan et al., 2007.

### The global context

Like many other environmental policy schemes, green labelling of electricity has its roots in the environmental awareness of the 1960s. But, as the reader will see later in this book, the question of what energy sources ought to be considered 'clean', 'green', and 'harmless to biodiversity and nature conservation' is the subject of more intense and animated debates than in many other sectors. For our purposes it was crucial to dig deeper into this extraordinarily conflict-laden sector, as we expected the translation from complexity to a categorical label to be particularly challenging (and therefore especially illuminating).

The energy crises in the mid-1970s and 1980s triggered efforts among governments, business, and NGOs to find and promote alternatives to fossil fuel-based electricity. Many perceived nuclear power – along with hydropower – as the only realistic alternatives, although nuclear accidents and exploitation of untouched rivers in the 1980s and 1990s have kept debates and environmental protests very much alive in all the regions of our study, and in most corners of the world. Over the last 15 years, the use of renewable and 'green' energy sources has attracted worldwide attention, based on worries about global climate change, local air quality, energy security, and possible exhaustion of fossil fuels (e.g., Ek, 2005; Gan et al., 2007). The rather cryptic term 'green electricity' includes energy generated from solar, wind, and wave power; certain biomass, low-impact, or early constructed hydropower plants; and geothermal power (Goldemberg, 2006). Increased promotion of green electricity has followed international treaties, including the United Nations Framework on Climate Change in Rio in 1992 and the Kyoto Protocol in 1997 as well as the International Action Programme at the International Renewable Energy Conference in Bonn (2004). In spite of these concerns and interests that all speak in favour of green electricity, it accounts merely for a marginal proportion of electricity generation in the world (Vachon & Menz, 2006).

Green electricity labelling can be found in several European countries as well as in North America, Japan, and Australia. Moreover, China is increasingly favouring voluntary green electricity schemes, as has been shown in Shanghai City in 2005. Voluntary green labelling of electricity is typically used as a supplement to obligatory regulations and quotas. A stated advantage of voluntary electricity schemes is that they may help consumers become more involved in broader implications of their electricity use. The flexibility – and the requirement for less government responsibility – is also a reason why several governments are in favour of voluntary electricity schemes (cf. Gan et al., 2007, p. 153).

## The United States: diverse levels of green empowerment across states

First of all, it should be mentioned that most of the US policy implementation concerning green electricity development is taking place at the state and local levels. Thus, only part of the American policies – the federal ones – can be compared on a same-level basis with policies within the EU's member countries. However, our aim is not to make full comparisons, but rather to indicate certain trends. On going through the policy research on US green electricity development, it is clear that several researchers perceive the United States as lagging behind, largely due to the lack of a clear national policy (see, e.g., Gan et al., 2007). Nevertheless, other researchers stress the broad range of green electricity policies across states within the United States, and thus highlight the innovative and progressive developments in certain states (see Vachon & Menz, 2006). Regardless of what view one takes, several initiatives and programmes have been developed, such as certification programmes, marketing and advertising guidelines in the United States to address the credibility of green power.

When the US Department of Energy provides information about how to purchase 'green power', it stresses the consumer challenge of knowing that 'they get what they pay for'. For instance, it is important to ensure that different organizations are not double-counting the same green power benefits. The abstract and invisible nature of green electricity makes such concerns particularly important to handle through certification schemes. In the United States, as elsewhere, third-party certification of green electricity sets standards in terms of overall environmental impact, minimum levels not just of renewable resources, but also of environmentally acceptable ones. Moreover, the ethical conduct of suppliers, including transparency and the avoidance of misleading information, is a task for the third-party certifiers. The Green-e programme is a leading independent third-party certification and verification programme in the United States.[28]

More broadly, it is important to note that voluntary schemes for green electricity are not merely a matter for NGOs and the private sector. For instance, a major coordinator of green electricity schemes is the Environmental Protection Agency's Green Power Partnership, which is part of the US government. Hundreds of organizations are currently voluntarily participating in the Green Power Partnership. The partnership offers expert technical advice to organizations that want to purchase green electricity; moreover, it claims to help organizations create credibility in the marketplace, ensuring stakeholders' confidence in the green power purchase of the organizations.[29]

From our point of view, what is interesting to note about the US green electricity programmes is how they allow consumers to choose not only 'green' electricity sources, but also specific ones (wind power, solar power, etc.) as well as the geography of these choices (Arizona-based, Texas-based, etc.; see the concluding chapter).

Finally, we should mention an important exception to the pattern of diverse eco-schemes of electricity across states in the United States. The most well-known green energy label in the United States is the Energy Star. In 2007, public awareness of this label exceeded 65 per cent of the population in that country.[30]

Energy efficiency – not the environmental impact of the energy sources – is the basis of this label. The US Environmental Protection Agency (US EPA) set it up in 1992 in cooperation with the US Department of Energy (EREN DOE). The participants or an independent testing laboratory test the products in order to see that the products meet the standards. Contrary to many other labelling schemes, there are no charges for participating in the Energy Star programme.

The US Energy Star covers a broad range of products: from office equipment to the buildings sector, lighting, consumer electronics, and residential heating and cooling equipment. A few years ago, critics used to point to the fact that a majority of the equipment in the above-mentioned product areas – up to 80 per cent – was meeting the Energy Star's efficiency standards. Recently, however, the scheme has reduced this arguably watered-down level to 25 per cent.

The Energy Star has become relevant far beyond the US borders. In the EU, the Commission established a European Board based on the Energy Star (EUESB) in 2001. The Board implements and promotes the programme. Countries such as Korea, Japan and Australia have also joined the Energy Star Programme.[31]

### The EU: mandatory efficiency label with rating

As mentioned above, there are a number of models for promoting renewable and 'green' electricity sources, not least in the European Union. Voluntary green labelling schemes in Europe have so far largely been a national affair. Different countries promote voluntary labelling to various degrees.

Although there is no common, voluntary green labelling scheme for green electricity in the EU, there are a couple of directives that constitute a basis for green labels on electricity in the member states.[32] However, there is one common European labelling scheme for energy that is omnipresent, largely because it is mandatory. By law, the EU's Energy Label must be shown on all washing machines, electric tumble-dryers, dishwashers, air conditioners, lamps and light bulbs, among other products. This label indicates the level of energy efficiency and not the energy sources used in, for instance, the production phase.[33]

In contrast to most other labelling schemes studied in this book, the European energy labelling has a rating scale based on energy efficiency, from 'A' (the most efficient) to 'G' (the least efficient). In addition, figures on the label express the energy use. In certain sectors, the industrial actors consider mandatory labelling as a stick rather than a carrot. When discussing the EU energy labelling, however, Bertoldi

(1999) maintains that it 'provides both a carrot and a stick, labelling good as well as inefficient products, so manufacturers and retailers have a twofold incentive to offer more energy efficient products'. Waide (1998, 2001) holds that average sales figures of energy-efficient home appliances within the EU have increased by 29 per cent. Moreover, a more recent study indicates a high willingness to pay for A-labelled products compared with C-labelled products (Sammer & Wüstenhagen, 2006). It is particularly interesting to note the extension of this rating scale. Since July 2004 A is not the highest rank. Instead, for certain appliances 'A' is divided into three categories (A, A+ and A++). Sceptics might worry that the label has become inflated, perhaps due to pressure from producers. And it would be relevant to ask why an adaptation to more efficient appliances could not be done by making the scale stricter, so that 'A' today would require higher efficiency than it did, say, five years ago. This case reveals some of the dynamics and dilemmas concerning the symbolic differentiation on which all green labelling relies.

### The national level in the EU: The example of Sweden

As to the national level, we have already mentioned that this is where one may mainly find voluntary labelling schemes, whether they are organized and controlled by NGOs, governments, companies, or hybrid organizations combining these. Among European countries, we examine in more depth the Swedish voluntary green labelling scheme indicating the use of green electricity sources.[34] An environmental movement organization, the Swedish Society for Nature Conservation (SSNC), controls the Swedish green labelling of electricity. The label is called 'Good Environmental Choice' (in Swedish, 'Bra miljöval'), which SSNC initiated in 1992 after some disappointment with the Nordic Swan.[35] Good Environmental Choice has developed into one of the most comprehensive green labelling arrangements worldwide and it is an exception that a single EMO controls the scheme (Rubik & Scholl, 2002). That is also an important reason why we wanted to include this labelling arrangement as a case (without focusing on all its product categories).

SSNC is one of Sweden's largest environmental organizations in terms of membership (Boström, 2007). All types of actors, including consumers, energy companies, and state authorities respect the 'environmental competence' of SSNC (Boström, 2001, 2007). The SSNC is generally considered an obvious stakeholder and discussion partner in the electricity and other sectors. The principal requirement is that only electricity generated through renewable energy sources can be labelled. This requirement sounds simple enough, but has turned out to be subject to substantial questioning and criticism, particularly in relation to the exploitation of rivers in Sweden (see later in this book). The same issue is a heated one in many corners of the world, including the United States and China.

## Green and ethical mutual funds (SRI funds)[36]

In research on consumer and investor policies, green and ethical mutual funds are not usually treated as green labelling (or eco-labelling). Still, SRI funds appear to fit our definition of green labelling quite well. And since SRI labelling is rarely studied in comparison with other sectors that use green labelling, we think that choosing this sector would help us to carry out a broader and more insightful analysis, which includes the financial world.

Decision-making concerning investments, not based solely on financial concerns but on a broader value basis, has been going on for millennia. Not surprisingly, the origins are largely religious. And it is no coincidence that it was a religious leader, Pope John Paul II, who wrote in Centesimus Annus, N. 36:

> Even the decision to invest in one place rather than another ... is always a moral and cultural choice. (in Beabout & Schmiesing, 2003, p. 63)

Yet, despite its long history, the academic and broader societal interest in what is nowadays referred to as 'socially responsible investment' (SRI) – of which green and ethical mutual funds are a part – dates back only 15 years (Sparkes, 2001, p. 196). Aside from its religious roots, SRI also had its origins in the 1940s, when government agencies and unions in some countries avoided investing in companies that were seen as unfair in their labour practices (Martin, 1986 in Hill et al., 2007). Like much environmental and social concern, the idea of taking such 'externalities' into account in investing grew stronger in the late 1960s and 1970s, in line with public reactions to the Vietnam War, social upheavals in urban areas, and environmental degradation (Spencer, 2001). Public concerns with global labour standards and human rights violations in production of goods and services on a globalized market have gone parallel with concerns for environmental problems during the entire life cycle of products. Social and environmental concerns among institutional and private investors have, of course, several overlaps, although they have resulted in partly separate indexes and mutual fund categories with different profiles. Today, it has become commonplace among SRI researchers to note an enormous increase in SRI activities, not least in the interest in green and ethical mutual funds among politically concerned household consumers along with institutional investors. Krumsick, for instance, claims that 'socially responsible investing ... is more widespread than ever, in the U.S. as well as Europe and Asia' (Krumsick, 2003, p. 583, in Hill et al., 2007). According to some figures, American SRI activities (in financial terms) increased by at least 258 per cent between 1995 and 2005 (Social Investment Forum, 2006, p. iv). With slightly different figures, others make the same main statement about the strength and increase of SRI. According to Laufer (2003) over one in eight US dollars managed by professional money managers goes to investments where social and environmental implications have been taken into account.

While saving the more detailed considerations of screenings of green mutual funds for later sections, we should immediately mention three fundamental types of screens that SRI fund managers use. Green and ethical mutual funds may be based on

- *negative screens*: the fund excludes companies that do not meet certain criteria;
- *positive screens*: the fund includes companies that meet certain criteria;
- *best-in-class screens*: the fund assesses companies in relation to the behaviour of other companies in the same sector. The fund chooses the ones demonstrating best practice.[37]

There are a number of organizations providing the principles and data for the three types of screens. At the international and general levels, there are the Principles for Responsible Investment (PRI), which provide a broad frame for SRI fund managers worldwide. In 2005, the United Nations Secretary-General invited a group of leading institutional investors from 12 countries to participate in the Investor Group, which develops principles for responsible investment. Until January 2006, the group had several face-to-face deliberations with a group of stakeholders from the investment industry, governmental and non-governmental organizations, as well as civil society and academia. The result was PRI. The United Nations Environment Programme Finance Initiative (UNEP FI) and the UN Global Compact coordinated the process.[38] Two other related principles and indexes worth mentioning are *FTSE4GOOD*, which is an index utilized for measuring the results of companies which meet globally acknowledged environmental and ethical standards,[39] and *The Dow Jones Sustainability Indexes*, which were launched in 1999 as 'the first global indexes tracking the financial performance of the leading sustainability-driven companies world wide'.[40] SAM was established in Zurich in 1995 as the first investment group focused exclusively on the integration of economic, environmental, and social criteria into investing.

It is fair to say that the United States has been a pioneer in the development of SRI funds, which is therefore a case that contrasts with the other US cases. To mention just a few American SRI organizations with a strong position, there is the Social Investment Forum (SIF), which is a member association based on more than 500 social investment practitioners and institutions, partly involved in mutual funds. The forum seeks to 'integrate economic, environmental, social and governance factors into their investment decisions'. In addition to this forum, there are a number of ethical and environmental consulting firms which help SRI fund managers select and exclude companies. Well-known SRI fund companies include Domini Social Investments LLC, Calvert Group, and Sierra Club Mutual Funds.

Even though the United States is often perceived as a pioneer concerning SRI funds, other individual countries have had a very strong development.[41] The United Kingdom, for instance, is often analysed in SRI studies (see, e.g., Friedman & Miles, 2001), and, like SIFs in several other European countries, the UK Social Investment Forum, UKSIF, has been instrumental in the growth of the SRI sector. Not surprisingly, the corresponding group in Sweden, Swesif, has been quite influential in promoting sustainable investment (Swesif, 2007). Unlike SRI fund companies in several countries, not least the United States and the United Kingdom, it is not typical for Swedish SRI fund companies to try to affect actively the environmental or social policies of the companies that they consider investing in. Three fund companies with such an ambition to affect companies, according to Skillius (2005), are Banco, Swedbank/Robur, and Folksam/KPA.

There are a number of organizations that provide principles and data to serve as the basis for all types of environmental and ethical screenings. Among the sectors analysed in this book, SRI funds are the least standardized type as well as the most multifaceted in terms of definition. In addition, SRI funds are the most privately organized kind. When comparing our cases the first impression is that the multiplicity of SRI funds – in terms of organization, screening processes, and low degree of standardization – makes this case analogous with the early stages of several other cases, such as organic food and forestry. There, producers also made a plethora of green and ethical claims, before regulations and standardizations of such claims were in place. SRI fund companies rest their criteria and assessments on broad international guidelines from the UN (see above), the International Labour Organization (ILO), the OECD, the Global Reporting Initiative (GRI), Amnesty Business Groups, and the like. Furthermore, the SRI fund companies often purchase the services of ethical analysis groups and consultants externally, and combine these services with internal environmental and ethical policies.

To be sure, 'the SRI community' has taken certain steps away from this multiplicity. The community has developed a voluntary quality standard of ethical analysis, supported by European analysis companies and by the EU. Moreover, an organization has been established to support the use of this standard. This is an example of auditing in the second order, where the auditors are checked, mainly in terms of transparency. The organization is called the Association of Independent Corporate Sustainability and Responsibility Research (AI CSRR).[42] Despite such steps towards partial standardization, it would be naïve to disregard the continued interests, among several stakeholders, in further developing green and ethical product and tariff *differentiation across companies*, something that might be particularly strong among SRI fund managers. We will discuss this in more depth later in this book.

## Paper labelling – the Swan versus Paper Profile[43]

Finally, we wanted to include a Swedish case indicating a clear green labelling backlash to obtain more information regarding labelling challenges, counter-arguments, and debates between proponents of different types of eco-standards. Such a case would also balance the other Swedish 'success stories', which would help us avoid giving an overly optimistic picture of the Swedish labelling situation. A particular case within the Nordic Swan arrangement fitted our purposes well.

As mentioned earlier, the Nordic Council of Ministers started the Nordic Swan, which was the world's first multinational eco-label (Micheletti, 2003). The Swan is a government-run eco-label, although both producers' organizations and SMOs (e.g., FoE and consumer organizations) participate in the organizational arrangement and in the standards development. It is made up of national committees – each with broad interest representation – and these committees are in turn coordinated by the Nordic Environmental Label Board. The Swan labels products in areas such as household chemicals, paper, office furniture and equipment, washing machines, textiles, DVD players, and hotels (Stø, 2002). It has its strongest market impact in Sweden, and it is the best-known eco-label in the Nordic countries.[44]

Our case refers to a situation in which business actors that had used a labelling scheme subsequently abandoned it. As part of their efforts to improve environmental performance and communication, Swedish paper producers joined the Swan in the early 1990s. In 2001, there were almost 100 licences for printing paper in the Nordic countries. However, after the revision of paper criteria for the Swan the same year, the producers chose not to renew their licences and they instead launched another eco-standard, named the Paper Profile (Nilsson, 2005). Only about 20 licences for Swan-labelled paper were renewed and the paper producers started to use Paper Profile, despite strong criticism from various stakeholders including certain professional buyers. This alternative eco-standard is a kind of standard for reports and declarations. In the Paper Profile, environmental information is presented according to a predefined form. The standard does not stipulate prescriptive requirements that need to be fulfilled. It defines a number of environmental parameters that must be mentioned in the form – largely the same parameters that appear in the paper criteria of the Swan, Good Environmental Choice, and the EU flower but without their specific threshold requirements. In contrast to these labelling programmes, an industry association with (Nordic) paper producers as member organizations runs Paper Profile. No other kind of stakeholder is involved.

Why did the paper producers leave the highly credible Swan and join Paper Profile? A number of arguments were put forward. Paper profile adherents expressed strong criticism of the Swan. For example, they complained about the inflexible, complicated, and inadequate criteria, as well as poor representation of paper producers in the organization. They expressed unwillingness to be controlled by a third party and thought that labelling had become a poor instrument for marketing and for environmental communication in general. They also stressed the need for international harmonization in a globalizing business, and a Nordic scheme did not fit well in that context. In addition to such arguments – further analysed in Chapter 7 – the policy context created opportunities for powerful counteraction despite the emergence of heavy criticism of Paper Profile. We do not go into detail in this chapter. In subsequent chapters, however, we will get back to this case to analyse certain arguments, policy conditions, and dilemmas in green labelling processes.

# 6
# Sceptical and Encouraging Arguments

Previous chapters should have indicated the vast spectrum of views concerning the advantages, disadvantages, benefits, and drawbacks of green labels. This chapter gives a systematic overview of the main arguments of scepticism and encouragement. The chief purpose of this chapter is to illustrate the multitude of arguments in labelling debates and thereby shed light on the challenging situation for labellers in their translation of complexity into categorical labels. Furthermore, another purpose of the chapter is to give the reader examples of the arguments, to be consulted for reference or to be read through as an ordinary chapter.

Organizing the arguments has not been an easy task, we must admit. Yet, the challenge that we have encountered in this process could be seen as a process where we have gained much knowledge useful for the rest of the book. *The first challenge* was that it turned out to be virtually impossible to make clear-cut distinctions between arguments based on *who* says what. The flexible use of arguments across SMOs, business actors, state actors, scientists, and consumers has been striking to us as analysts, and we give several examples of this throughout the book. On the one hand, actors such as EMOs may be more likely to address some of the arguments described in this chapter. On the other hand, disagreements among EMOs are common and sometimes heated. We observe similar heterogeneity also among all other actor categories. In addition, labelling projects see many unexpected coalitions across such actor categories as SMOs and business actors (see Chapter 9). Consequently, this chapter tries to strip the arguments of their subjects (although we will exemplify specific arguments with quotations and references for the sake of good illustration). The chapter gives an overview of the arguments without analysing who uses what argument and when. Thus, we try to introduce parts of the argumentative toolbox, which subsequent chapters provide with subjects, policy contexts, organizational processes, and framings.

*The second challenge* of trying to structure the arguments was that it turned out to be virtually impossible to make any clear-cut distinctions at all. Yet, we try again to turn this more profound analytical problem into a lesson about the field. It highlights the basic position presented in the first chapter, namely that labelling takes place *between* facts and values, science and politics, *between* actors and institutions, in ways that do not allow the analyst to make traditional dualist separations. This chapter uses a distinction which, to be sure, does not escape overlaps, but is based on how the actors in our case studies conceive of labelling. It is therefore important to stress once again that the arguments – and the underlying epistemic or power-related positions – are not our own. Instead, they derive from our comprehensive analysis of data from interviews, documents, and so forth. The distinction stems from the following notions in our empirical material:

Firstly, labelling largely takes place – and partly challenges – principles in *the market realm*, principles such as fair competition between equal products, and the avoidance of misleading or irrelevant information. Secondly, labelling is produced with the aim of giving *knowledge and information*, particularly about possible environmental and health-related impacts. Thirdly, going beyond the market realm, labelling involves ideas about *green governance* in general, for instance, how previously separated actors, organizations, and realms should all participate in green policymaking.

We are not completely neutral in relation to the arguments, either from a theoretical or descriptive point of view or from a normative one. Some epistemological assumptions made in our introductory chapter and in the normative discussion in our concluding chapter would prioritize certain arguments before others. However, in this chapter, we treat the arguments as empirical social constructs, and in this chapter we do not push our own positions.

## Market-oriented arguments

### Encouraging arguments

*Labelling empowers consumers.* This is maybe the most straightforward argument, and maybe the most common, and therefore relevant to start with. Although it appears intuitively simple and plain, one should note that two theoretical arguments are involved here, explicitly or implicitly.

The first is that labelling is good since it (supposedly) does something good for externalities, that is, the environment, reduced social harm, and so forth; this is the argument about *an extrinsic value of political consumerism through labelling*. The other argument is that buying eco- or fair-trade-labelled products helps consumers express their political identities. Rather than a concern for externalities, the latter argument implies that political consumerism through labelling has an *intrinsic value* (Klintman, 2006). In labelling there are concrete channels for consumers to do something proactively, individually, responsibly with various non-economic goals in mind, without having to rely on existing collectivities, such as political parties that they may have difficulties identifying with. Moreover, ecological or socially friendly shopping can be a way of expressing religious or ethical sentiments. It can also be a way for individuals to express a particular social status position.

Intimately related to the intrinsic value of consumer empowerment is the claim that consumer information can never be harmful: 'Since people want information through labelling, just let them have it!' This is an argument that easily fits with a multiple set of – even opposite – ideologies and framings, and which has therefore been instrumental in conflict resolution and frame bridging. It is a message that may resonate with storylines about 'economic man', 'rational choice', 'transparency', 'consumer democracy', 'consumer empowerment', and so on. It is indeed difficult to challenge arguments about people's intrinsic right to information, especially concerning credence goods whose qualities are impossible to evaluate when one is using them (Darby & Karni, 1973). This position may frame as paternalistic the objection about labelling providing consumers with irrelevant distinctions. This 'consumers' right to know' frame has helped resolve the largest controversy about green labelling that has taken place so far: the Big Three controversy in the United States (see Chapter 8; Klintman, 2006).

*Labelling creates new business opportunities.* Anyone taking an interest in market dynamics and opportunities for new niche markets may be impressed by the exponential growth of certain ecological and fair-trade-oriented market niches in several countries. From that point of view, it can be perceived as positive that labelling creates price premiums, market opportunities, and entries to the market. Firms may see labelling as an opportunity for market entry, and for creating a green niche in competition with firms dominating the market. The head of the eco-labelling programme 'Good Environmental Choice' run by the SSNC (also the Swedish label for green electricity) mentions a striking example in the Swedish chemical engineering industry

(Boström, 1999). When SSNC introduced their label for chemical engineering products in the late 1980s, market niches were opened for producers that had previously had difficulty in competing. A strong oligopoly with four large producers was broken, and in the early 1990s this market had hundreds of producers. In terms of profit and premiums, studies collected by the Organic Trade Association in the United States indicate substantial price premiums on several organically labelled food products, both domestically and on products exported to Europe and Asia.[45] Actors use such reporting of concrete examples as a powerful argumentative weapon in debates about the greening of business.

*Labelling stimulates a green image of progressive companies.* According to this stance, labelling effectively contributes to greening the image of certain companies, something that is often good if the company 'deserves' it. From this viewpoint, green labelling is seen as a *credible* way to express greenness and corporate responsibility. It is credible because of the great involvement of highly trusted NGOs and SMOs. The implicit assumption is that the involvement of such external stakeholders automatically guarantees good standards. Moreover, labelling implies a symbolic differentiation between 'good' and 'bad' companies and products. Being categorized among the 'good' ones can be a way of gaining competitive advantage.

This benefit of labelling is closely connected to the 'reverse' claim, namely that labelling can help companies by preventing or dealing with negative publicity. Businesses may fear consumer boycott campaigning staged by media-conscious SMOs, which may result in reputational damage, especially in times when brands, symbols, and messages are increasingly important corporate assets (cf. Chapter 2); and labelling can in this context be offered as a method for reputational risk prevention.

*Labelling stimulates a green image of progressive industry sectors.* Labelling can also be seen in a broader light, namely in terms of how it might affect entire industry sectors – not merely single companies. A statement sometimes heard in greening-of-business debates across actors is that labelling and certification actually may contribute to a green image for a whole industry. For instance, in the case of Swedish seafood labelling, some spokespersons of the fishing industry claimed that introducing a labelling scheme would be beneficial for the whole fishing sector, despite the fact that only a minority of fishing companies will be able to get a certificate. Their rationale for this position is that the whole industry has received so much public criticism in recent years (due to over-fishing, a tendency to cheat with legal requirements, etc.) that just being able to visualize a few 'good examples' within the industry would benefit all actors in the industry.

## Sceptical arguments

*Labelling is inefficient because it cannot scale up.* This sceptical argument, which we mention very briefly here because we discuss it in both the introductory and concluding chapters, relates to a sceptical view of consumers' incentives for free-riding and 'low willingness to pay'.

*Labelling provides the market with misleading separations of identical products.* It is possible to claim that labelling of products that are not essentially different, instead of increasing consumer choice, actually *reduces* consumers' freedom of choice. As a consequence, labelling is a market-distorting mechanism that illegitimately reduces the scope of the market. For example, it has been held that labels stating that a certain food product is 'GM-free' may stimulate irrelevant considerations among consumers and erroneously imply that GM food is inferior to conventional food (Klintman, 2002b). In several other cases sceptics of labelling accuse the schemes of being misleading, cases where the labelled end products do not differ from 'conventional' products. As to green electricity labelling, a major challenge to consumer motivation appears to be that the end product is identical to the conventional product. The green electricity consumer gets the same electricity as those who have not made any consciously green choice of electricity. In interviews, several actors indicate that this lack of visibility through concrete product separation is related to people's mistrust (Lindén & Klintman, 2003). If the difference is primarily administrative (with a different tariff that goes into renewable sources) rather than physical (which would entail that electricity from alternative electricity sources would be in our grid) the green label is insufficient for visualizing and motivating a consciously green electricity choice.

*Labelling creates unfair advantages for big businesses.* An often-mentioned aspect of labelling and certification schemes is their financial costs. Huge costs are of course a negative argument in themselves; but the sceptical attitude relates also to unfortunate distributional side effects of the costs. In several sectors and countries, there is a claim that only big businesses can afford to participate in a certification programme. The argument implies that such programmes and labels sometimes reward the wrong players, who can pay to become a member of the programme, and have the technical capacity, management structures, and skills to implement standard requirements and to market their products. We will analyse this in Chapter 9.

*Labelling stimulates an inflation of green claims.* Labelling may also be rejected because it is seen to have failed in the ambition to single out 'one best choice'. In a way, this is the opposite of several other objections. If it becomes too easy to achieve green labelling criteria, several of the arguments *in favour of* labelling cease to be valid. The label becomes a symbol of the 'ordinary' and 'average' rather than the 'best', a criticism that has been levelled at the EU energy label (A++; see previous chapter) and Nordic Swan regarding some product categories (Nilsson, 2005). As a response, corporations may use other eco-standards or develop other tools to express their greenness.

*Labelling disturbs the continuity of business plans.* Revisions of labelling criteria are often done repeatedly in order to comply with the principle of 'continuous improvement'. Such continuous revisions create a degree of unpredictability, which certain actors dislike. Each revision of a certificate constitutes a risk because the certified company may not be able to attain the new standard requirements. Standards of procedures, such as environmental management systems (e.g., ISO 14000), also include notions of 'continuous improvement', but in these systems certified companies are not required to comply with fixed rules and thresholds that an independent organization has set. Instead, the companies themselves set the goals and accordingly design the long-term process. For example, the system of performance-based standards and fixed, repeatedly revised rules and thresholds does not fit – it is claimed – an industry with heavy physical investments, such as the pulp and paper industry (Nilsson, 2005). Long-term thinking and continuity have to be the basis for such planning. The calculated pay-off from a new investment is based on a much longer time horizon than are the rather sudden labelling revisions. From this perspective, even the traditional command-and-control mode of regulation provides better continuity and discretion than labelling (Nilsson, 2005, p. 27). Environmental declarations, such as the Paper Profile, are frequently mentioned as an alternative.

*Labelling is an instrument of disguised protectionism of the North.* A concern from the perspective of global free trade advocacy is that labelling may hamper free trade. This is a position used in combination with the emphasis of WTO principles concerning technical barriers to trade. The debates have been intense in relation to green labelling of seafood and forest products. In the United States, this has been one of the main principles behind the FDA's decision not to require that GM foods be given a mandatory label. We discuss this issue more in the next chapter.

*Labelling is about marketing rather than about reduction of external harm.*
Another argument is that labelling has shifted from being a tool for
social and environmental change to simply become a marketing tool,
which is heard frequently among certain organizations of small-scale
organic farmers as well as among certain researchers on organic food
production and markets. They stress that organic food production is
increasingly trapped in the conventional market logic, dependent on
large-scale systems and on excessive compromises of organic principles, and so forth.[46] Ironically, also an 'anti-organic' position
('pro-conventional or GM industry') may claim that organic labelling is merely a marketing tool. From an 'anti-organic' position the
claim that organic labelling is only a marketing tool may imply that
the organic label is arbitrary in that it does not help to distinguish
between anything substantial. For instance, Alex Avery, director of
research at the Hudson Institute Center for Global Food Issues, states
that 'The label ... is not ... better ... in any way. It is purely a marketing
label' (the statements by the USDA *in favour of organic labelling* have
been nearly identical; see McAvoy (2000), note 61 above).

## Knowledge-oriented arguments about environment and health

### Encouraging arguments

*Labelling helps consumers and professional buyers get the most concise type of
information.* This is of course one of the most basic and obvious reason
for establishing labels in the first place. In today's complex modern
societies, people simply lack sufficient knowledge to deal with various
problems. The cooked fish on the plate does not inform the consumer
that it once lived in a healthy stock. People are dependent on expert
systems (Giddens, 1990). Even 'professional' buyers need advice. On
certain occasions, they may have a sufficient knowledge base to employ
alternative tools such as 'environmental declarations'. They may also be
able to evaluate different parameters, such as relative energy use or
emissions of carbon dioxide. Yet, professional buyers may have responsibilities for the purchasing of 5000 different products, and it is just
impossible to interpret and assess all details in the declarations, and to
compare different buying options according to the available 'neutral'
information. They do not want just any 'neutral' information (Nilsson,
2005). They demand evaluations from credible experts, evaluations
that the green label provides. They want someone to trust.

*Labelling criteria assist in informing about sustainable practices.* 'What is sustainability in translation to our practice?' Authorities, the public, NGOs, or companies may for various reasons require assistance in reforming (or informing about reforming) practices, but they do not know how to do it. They may commit to sustainability principles and triple bottom lines reasoning, but may lack the knowledge, routines, and tools to choose the best strategies. Such knowledge may be embedded and codified within the labelling programme as such (cf. Jacobsson, 2000). The labelling instrument may offer a practical tool to solve or deal with complex problems that have to be solved or dealt with anyway. For example, this argument appeared in the Swedish FSC case among both companies and policymakers. Forest companies felt various kinds of pressures (legal, economic, social movement) to reform forest practices but they simply lacked adequate expertise and tools; they found these within the FSC standards.

*Labelling stimulates the production of new knowledge.* A great number of informants from our case studies testify that labelling processes stimulate dialogue and mutual learning among actors. For many of them this seems to be the experience gained from the labelling process, rather than an initial encouraging argument. However, a few informants have also turned this scenario about learning and knowledge production into a general *ex ante* argument. As labelling is dependent on the experience and knowledge of a great number of actors, it can contribute effectively to the acquisition of new knowledge or to a new synthesis of previous experiences and knowledge. For example, in the case of Swedish seafood labelling, certain informants believed that labelling could be a way of moving ahead from a background of stalemate and polarized positions.

### Sceptical arguments

*Labelling is pseudo-science.* This argument contends that it is impossible to prove – in an absolute sense – that labelled products are environmentally friendlier, fairer, healthier, safer, and so forth, than conventional products. Labelling processes may be seen as *exclusively* a part of political struggles among actors. According to this view, the criteria and thresholds stated in the standards primarily – or only – reflect ideological viewpoints. This is a common position among actors that criticize organic food labelling and proposals of mandatory GM labelling, for instance in the United States (Klintman, 2002a, b; Klintman & Boström, 2004). This kind of objection tends to be based on *judgemental relativism*, an epistemic position claiming that all knowledge, including information systems such as labelling, is completely socially determined and thus impossible to compare or assess the value of (Bhaskar, 1989). Interestingly, those who say NO (and thus express judgemental relativism in their rejection of labelling) can simultaneously embody the opposite position – *epistemic absolutism* (the claim that knowledge can be a direct and complete reflector of the truth) – regarding the safety of the technology as such (e.g., GM technology) that they do not want to label (Klintman, 2002a, b). Such 'crossovers' in the reasoning are less likely in the European context where the reliance on a 'precautionary principle' (despite the ambiguity of the term) is much more prominent.

*Labelling does not take environmental and social consequences sufficiently into account.* Even if commentators reject the endpoles of judgemental relativism and epistemic absolutism, they may argue from a methodological point of view that the labelling criteria do not reflect the best moves towards sustainability. They argue that the labelling criteria are too rigid, inadequate or unable to adapt to changing conditions. Analysts may maintain that a particular practice, which is subject to standardization, varies to such an extent that accurate environmental measures, within the context of labelling, are impossible to define (Nilsson, 2005). Instead, flexible and adaptive management is seen as more scientifically adequate. For instance, researchers may claim that organic farming is even more damaging to the environment than some forms of conventional farming that use pesticides restrictively as in, for instance, so-called 'integrated farming' (Trewavas, 2001). By using a small amount of pesticides, the farming can reduce the mechanical prevention of weeds. This in turn allows for reduced use of tractors and thereby also for reduced discharge of carbon dioxide. An optimal balance between various environmental goals and criteria would need approaches less rigid and more sensitive to shifting conditions. Likewise, labelling criteria need to be standardized and fixed within a geographical area, which actually has great variation in terms of eco-systems. For example, in the Swedish FSC standard there is a criterion that 5 per cent of the landowner's forests – the area with greatest nature value – must be removed permanently from exploitation, which has been seen as more or less accurate because of the greatly varying landscapes in such an elongated country as Sweden (Boström, 2002).

*Labelling allows only for limited traceability.* It is often heard in the debates that labelling just does not work. The objectives of labelling might be good, but labelling is simply not feasible in practice. There can be hundreds of arguments behind such objections. One example relates to the chain-of-custody arrangements, which labelling arrangements normally require. Objectors may claim that it is impossible to track raw material to the ultimate source for a composite product because a company may rely on hundreds of subcontractors. An American forest company executive explained:

> Logistically, it is extremely difficult to figure out where all of that wood comes from and then go out and try to see that all of these dozens if not hundreds of small private landowners are practicing FSC standard of forestry. Even if you really wanted to do it and thought that there was a reason for it – and I don't – logistically it would be almost impossible. (quoted in Cashore et al., 2004, p. 100)

Here, the challenges of substantive traceability (based on the substance of the products and processes, raw materials, ingredients, additives, pesticides, etc.) go hand in hand with the challenges of procedural traceability (based on the chain of actors, companies, and other actors involved). These two types of traceability closely connect with transparency challenges, a topic we return to in Chapter 10.

## Green governance-oriented arguments

### Encouraging arguments

*Labels are more credible than other standards.* From this perspective, labelling is a credible channel for expressing greenness in competition with other eco-standards such as environmental management systems (EMS) and environmental declarations. Well-developed forms of green labelling may accordingly have the potential of becoming both *legitimate* and *credible*, particularly in cases where labelling is based on independent, third-party standard-setting and certification. The argument is fuelled by concerns about other eco-standards such as EMS because such standards allow firms to define their own goals. There is widespread concern that these standards will not really make a difference in terms of environmental performance and that they may be used merely for legitimizing business as usual and to head off more stringent regulation (Clapp, 2005).

   *Labelling is part of a promising regulatory Plan B or promising regulatory supplement.* In texts of environmental regulation and so-called 'new modes of governance', labelling (along with other instruments of soft regulation) is typically portrayed as an alternative way, a 'second-best' strategy, for solving environmental problems, against a background of perceived regulatory voids or failures. Accordingly, traditional, state-led authoritarian regulation has repeatedly proved incapable of dealing adequately with environmental issues; given the greening of the industry trend it is reasonable to develop more market-based strategies. Claims of regulatory failures and regulatory vacuum were especially common concerning transnational forest and marine certification and stimulated the search for new market-based approaches. 'You can't just sit back and wait for governments to agree, because this could take forever', a WWF leader commented on the forest case (quotation in Bendell & Murphy, 2000, p. 69). Another example is electricity labelling, which developed partly as a reaction to the failure to phase out nuclear power. One reason for deregulating the electricity sector in Sweden and certain other European countries was, some actors have claimed, that it would create a product and tariff differentiation entailing, for instance, that green electricity labelling could be distinguished from 'conventional' electricity sources. For various reasons – moral, image-related, and so forth – a significant number of consumers and companies would accordingly partly prefer the green electricity label, and thereby help to solve environmental problems (Lindén & Klintman, 2003). In this vein, the leader of the Swedish Campaign against Nuclear Power (Folkkampanjen mot kärnkraft), Gunnar Landborn, claimed:

> A key to the phase-out [of nuclear power] is the deregulation of the electricity market ... People will be able [after the deregulation] to choose green electricity in the same way as they buy unbleached paper. With the constantly growing environmental opinion, and the new generation's conviction that change begins among ourselves, the green-minded consumers may rearrange the energy market more quickly than we may imagine.

A less categorical claim than the one about a promising plan B is that labelling is useful for *supplementing* existing rules. In the idea of labelling as a supplement lies the view that existing regulatory authorities have indeed achieved a great deal; at the same time, traditional regulatory arrangements and instruments cannot tackle all problems. These are similar arguments to the plan-B argument, but given a more collaborative frame. Concerning green-labelled electricity, a spokesperson at SSNC stresses the co-dependence of various instruments for green adaptation:

> Eco-labelling will not solve the problem of dreadful hydroplants in poor countries, or investments in nuclear power ... Taxes and laws

may get such things. It won't work with only strict regulation or *only* electricity with a green label …. Eco-labelling is a sort of lamp that gives light a bit ahead. It says: Look here, this works!

Similar pragmatic reasoning is found among actors from various organizations in all our cases. Some argue that it makes authorities able to allocate their resources for other purposes, in both policy formulation and inspection. Yet, this presupposes that authorities trust (possibly excessively) that certified businesses perform well, and that 'independent' certification bodies effectively audit companies.

*Labelling is part of a broadening of power in society, which is beneficial to democracy.* Labelling tends to involve non-state actors, including SMOs, often to a much larger extent than in traditional command-and-control regulation. This broadening of power can be perceived as a democratic advantage of labelling and other instruments of soft regulation. Several labelling schemes and administrations provide EMOs and other SMOs with significant formal decision-making power, which is a topic we discuss in Chapter 9.

*Labelling is politically efficient due to its collaborative nature.* Even *within* claims in favour of alternatives to top-down regulation, divergent views can be found about what 'sub-alternatives' are the most efficient ones. For instance, a distinction can be made between confrontational and collaborative strategies. Arguments in favour of labelling contend that it is a more effective strategy than other, more radical, strategies, for instance boycott strategies. To be sure, many politically engaged actors conceive these two strategies as complementary. Still, there are disagreements about which of the two ways is more efficient when working for social change. For example, in the FSC case, WWF questioned the effectiveness of tropical timber boycotts that were used in the 1980s, claiming that the boycotts did not lead to a reduction in deforestation rates (Domask, 2003). Boycotting led only to a devaluation of the standing timber in tropical forests around the world, which in many cases actually caused increases in deforestation as the forests were cleared for other uses such as farming and grazing.

*Labelling helps to develop friendlier relations across groups.* There was a similar encouraging knowledge argument presented above. Here, the potential to develop friendlier relationships among business actors, state authorities, social movements, consumer groups, and personnel within companies is lifted to an *ex ante* argument. A few informants have expressed a hope that deep tensions within an industry and among its external stakeholders should be resolved through a labelling process.

For instance, the FSC working process in Sweden that led to the implementation of the FSC label was supported by leading state officials because it was seen as an opportunity for interest groups to reconcile and to break the earlier stalemate in forest politics and policymaking.

## Sceptical arguments

*Labelling adds another layer of rules to an already overregulated industry.* In several of our cases, interviewees have maintained that 'we have a long tradition in our industry of acting in an environmentally responsible manner; we know how to do it.' From this perspective additional rules are simply unwelcome. This is a topic we expand on in Chapter 9.

*Labelling is based on excessive power of external stakeholders.* Some think a strong role for external stakeholders is good for democracy (see above in Encouraging arguments). Others are concerned. The argument essentially reveals an underlying power struggle in labelling. This struggle concerns which groups of actors should make decisions about core corporate practices. It also concerns which actors should have the various positions in the regulatory space (Hancher & Moran, 1989). Corporations, this argument implies, should decide themselves about their own conduct. Certain companies may have good relations with public authorities, whereas they consider it illegitimate the kind of involvement of external stakeholders (e.g., NGOs, SMOs) that is common in labelling. They argue that companies must have the sole responsibility for implementing measures in order to solve various problems – including problems with environmental and social consequences – since the *companies* bear the implementation costs.

*Labelling is often based on special interests of private rule-setters.* Private rule-setters are private, tautologically speaking. For example, Scandinavian authorities pushed such arguments when they rejected the MSC initiative. Informants from the National Board of Fisheries said that their main concern was the central role of a transnational corporation – Unilever – figuring as the main sponsor: 'I think that the MSC is just another consulting firm with the only aim being to earn money, completely following its own self-interest', one official said (our translation). They also thought that the MSC was not transparent. When they asked the MSC for detailed information they were told that it was classified and they were not welcome as participants. Hence, they thought they were denied any possibility to oversee that the process adequately took into account all interests in a balanced way. In the case of organic labelling in the United States, concerns within the USDA that

NGO-controlled labelling would jeopardize, for example, principles of nationalized standardization, led the US federal government to take control of the standardization of organic criteria.

*Labelling processes disregard the experience of public authorities.* Existing national and transnational authorities may have long traditions of dealing with regulatory issues. In this light, authorities may perceive it as disturbing that private actors try to occupy positions in the regulatory space. In certain labelling and certification schemes, initiators make a 'diagnostic framing' (of problem identification and attributions, see Snow & Benford, 1988) that traditional regulation (and regulators) had been a complete failure. This framing is often manifest in the fact that non-state initiators do not invite traditional authorities to take part in their rule-making activities. The 'prognostic framing' (ibid.) often contends that private or NGO-led 'soft' regulation, for instance labelling and certification, is the best way of resolving global crises over and above the jurisdiction of nation states (cf. Constance & Bonanno, 2000, p. 134). Criticism of this is often that private rule-setters do not respect existing authorities, their expertise, and their regulatory frameworks, and that the private rule-setters thus run the risk of throwing out the baby with the bathwater.

*Labelling provides only a shallow transparency.* Another concern is that the labelling schemes provide too little information. In this view, more should be revealed and transparent about what is 'behind' different products; consumers and other actors should not have to hold a simple trust in the categorical statements of labelling schemes (Klintman & Boström, 2008). Environmental declarations are often marketed with arguments that they include more detailed, substantive information about various aspects of the production processes (Nilsson, 2005). Preferences for more comprehensive declarations than labels are often based on the claim that declarations are better in line with the development whereby expert knowledge is growing and spreading among broad groups of the population in modern societies. In such societies, consumers – many of whom actually are professional buyers in big organizations – are knowledgeable and reflective. 'People need to know more' is a common statement in this context.

*Labelling implies an excessive consumer responsibility.* It is sometimes argued that it is the authorities and corporations that created the environmental, social, and animal-related problems in the first place. Consequently, one may hold that it is the authorities and corporations that should take the responsibility and bear the costs of solving them.

Moreover, these hugely powerful actors (corporations and authorities) are arguably the only ones with the true capacity to solve such complex problems. What can individual consumers do, particularly in light of the competition between various needs, tasks, and goals in their daily lives? Individuals may be unable to see their individual agency as part of a collective action, that is, as part of a 'we', of an 'imagined community of active co-consumers' (Halkier, 2004, p. 235) that helps to make individual consumer practices meaningful in relation to social change. The view also fits the classical problem of the 'free rider' (Olson, 1965), denoting that a person may enjoy the benefits of public goods that are supplied by other collective actors, while doubting that his or her individual contribution to the common good would make a difference to the continuous delivery of this good. From certain perspectives of economics it can be argued that individual end consumers should not have the responsibility and bear the cost of solving various 'external' problems. A related claim is that the costs of producing public goods, such as a clean environment, are only transferred to the end consumers. To set up and run a labelling arrangement requires additional resources, and someone has to pay: 'the poor, who must buy at the bottom of the market regardless of their personal opinions, pay a disproportionately higher share of the increased cost to the benefit of no one, especially themselves' (Alan McHughen, as quoted in Klintman, 2002b, p. 75).

*Labelling is not a radical enough instrument.* In debates about negative impacts of consumerism, it is sometimes maintained that labelling as a strategy is not radical enough, and that labelling even legitimizes unsustainable corporate practices. What is the core of this argument? Various actors and organizations claim that labelling is a poor way of expressing responsibility and accountability; it cannot tackle the most salient issues. Accordingly, labelling relies on severe concessions to big businesses – often with transnational corporations. Hence, labelling is unable to alter global relations of power asymmetry. The argument contends that moderate pro-labelling SMOs, such as the WWF, naïvely trust the businesses with which they cooperate. For instance, one leader from Survival International comments critically on the FSC label:

> Consumers can consume even more, companies can make profits, forest communities can make an income, the environment is saved ... No one and nothing is criticised. The causes of rainforest destruction and the invasion of tribal peoples' lands are not addressed. This is not a panacea, a placebo or even a quick fix, it is just slow poison. (quotation in Bendell & Murphy, 2000, p. 74)

Some of these opponents may refer to the 'small is beautiful' of the more radical environmentalism of the 1970s, and they notice that labelling clearly does not represent small-scale economic activities. According to this position, an ambition of all green strategies should be to restrict the power of large corporations and facilitate economic practices based on civil society associations (cf. Guthman, 2004).

## Conclusion

The debate within and around green labelling is not only a debate about its pros and cons. Important debates occur across the different encouraging arguments. These are not necessarily compatible. A great array of encouraging reasoning contains many win-win scenarios, which should be a good condition for the framing, introduction, and implementation of labelling. Yet, if actors want labelling for different reasons, stalemate is likely to occur. For example, whereas EMOs may see labelling as a promising strategy for improving the environment (since other actors and institutions have failed to take on their responsibility), business actors may primarily view green labelling as a promising way to improve their contact with governments, or to better their green image and public relations. Some proponents call for very strict standards that reflect 'top practice', whereas others hold that an inclusive standard is better, since the latter allows for a larger part of an industry to move towards sustainability by means of the labelling scheme. Certain actors in the United States have even advocated broadening the organic label so that it may also include 'naturally grown' GM food, which has led to hullabaloo among organic constituencies (Klintman, 2002a, b). Thus, different motives are far from easy to unite in the labelling process, and this may cause tensions in the design of labelling arrangements.

   When groups agree to start to initiate a labelling programme, the sceptical arguments do not simply cease to exist. They may particularly remain outside the labelling coalition, but they may also take shape within the coalition. The very awareness of the sceptical arguments can affect the debates about the goals, concerns, principles, criteria, and methods of labelling. Sceptics may start to encourage labelling, while remaining hesitant and without revising their basic concerns; counter-arguments can remain influential. In sum, the multitude of motives, arguments, counter-motives and counter-arguments constitutes a great challenge for framing and organizing processes as regards the possibility of creating a coherent, simple, and categorical label.

# 7

# Policy Contexts and Labelling

We would not expect to find many green labels in North Korea (although we have not been there). Perhaps it is also difficult to find eco- and fair-trade-labelled products in stores in the Dominican Republic, but it should be less difficult to find fair-trade-certified farmers in the countryside. However, it is far from self-evident that these farmers actually know that they operate under a fair trade labelling system (Getz & Shreck, 2006). It is probably easier to find green labels in Italy than in Hungary, easier in Norway than in Italy, and easier in Sweden than in Norway. On average, Swedish citizens may be more prepared to buy labelled products than citizens of the United States. Yet, American citizens who buy labelled products may be more ideologically committed to their purchasing behaviour than their Swedish counterparts. And debates about labels in the United States may cover more themes (e.g., survival of small-scale, local production) than such debates in Sweden. UK citizens may go to supermarkets such as Marks & Spencer to buy organic produce while Irish citizens may go to the farmers' markets to buy their ecologically sound vegetables (Moore, 2006).[47]

Labelling initiatives, arrangements, and debates appear to be tangled and shaped by existing patterns in different countries. In this chapter, we analyse a set of factors in order to see how policy context matters. We pay particular attention to Swedish and American examples, within our selected sectors, and we aim to discuss various ways in which policy contexts facilitate or obstruct the introduction and implementation of green labelling. We are particularly curious about why the tools of green consumerism appear to be easier to implement in certain countries and certain sectors, while implementation is more challenging in others. Why, for example, are there so many internationally recognized labelling initiatives in such a state-centred political culture as Sweden (see also Micheletti, 2003) compared with, for instance, the situation in the United States, where one could expect to see more of such active consumer policies, due to its distinctive market-liberal and consumer-oriented political culture? We are also interested in the ways in which the policy context affects debates and reflections on green labelling. These questions relate to our general themes concerning the roles of politics, trust, symbolic differentiation, and the relation between production and consumption. We should immediately acknowledge that no definite answer to any of these questions can be given. Our aim is to shed light on how policy context *can* matter, and that it can matter in many different ways; and that labelling can work despite varying conditions.

In the next section we will briefly theorize policy context and present a model that guides our analysis and discussion in the rest of the chapter. As the reader will see, we refer to four general context factors when analysing the role of policy context.

## Theorizing policy context

Figure 7.1 below illustrates our thinking of the policy context. We broadly distinguish four context factors: political culture, existing rules and regulations, organizational landscape, and materiality and technology. By using the term 'context' we refer to both 'cultural' (widely shared ideas, traditions, beliefs) and 'structural' (regulations, organizational constellations, physical constructions) elements. This provides opportunities for, as well as limitations on, action; furthermore, contexts shape processes of framing and organizing, as we shall see in subsequent chapters. It is important to note that the term 'context' may appear to indicate something external, exogenously given, to which actors passively have to adapt. Our understanding of the relationship between context, action, and process is more dynamic and open-ended. Although context is normally seen as something external to action, it is important to emphasize that actors may respond creatively to cultural (cf. Swidler, 1986) or structural (Giddens, 1984) elements in the context. For example, 'free trade rules' (described further below in the chapter) may appear very compelling or disturbing in the view of labelling agents, as free trade rules clearly affect the conditions for labelling in various ways. Yet, labelling agents can make creative use of such a context element. They can re-interpret them, circumvent them, and give them new meaning in the specific project. The bowed arrows in Figure 7.1 illustrate actors' potentially creative relation to context.

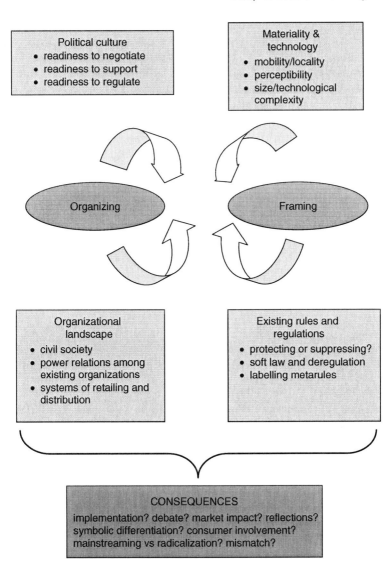

*Figure 7.1* Policy contexts and labelling

As we see it, the context elements can have varying consequences and their causal powers can work in opposite directions and neutralize each other. It is therefore strictly impossible to determine exactly to what extent a particular context element affects, for example, the successful introduction of a label; yet systematic comparison can reveal patterns that allow for some interesting interpretations. Different elements can facilitate labelling in one way (e.g., effective implementation) and obstruct it in another sense (e.g., hinder frame reflection and public debate). With our model and discussion we do not aim to present an exhaustive list of context elements that affect labelling.[48] Our basic aim is to argue for the relevance of understanding the role of policy context, not to provide the full story of how context affects labelling.

The four general context factors will be discussed in the rest of the chapter. Admittedly, the distinction between these four broad factors is not clear-cut when it comes to particular context elements. For example, ideas, assumptions, and beliefs within a political culture can be materialized and institutionalized in routines, rule systems, and organizations. The free trade ideal, for instance, may be seen as a cultural element at the same time as it is materialized in rules and embodied by organizations, such as WTO. For reasons of simplicity (the organization of the chapter), we present a particular context element under one particular factor.

## Political culture

The first general context factor that guides our analysis is political culture. Scholars in sociology and political science use political culture in comparative research on policy processes.[49] The concept of political culture generally denotes institutionally sanctioned modes of action as well as ideas, beliefs, and unwritten codes and practices concerning such diverse aspects as who the legitimate policy actors are (e.g., state *vis-à-vis* non-state actors); regulatory style (e.g., legalism vs. pragmatism); where (in which arenas, on what levels) policymaking should occur; the relation between science and politics; and the allocation of market-versus state-centred instruments (e.g., Christensen & Peters, 1999; Jasanoff, 2005). Furthermore, North-Western Europe and the United States are often used respectively as typical examples of a political culture characterized by consensus (in North-Western Europe) and conflict (in the United States). This separation of consensual Europe and adversarial United States is made more distinct through the typical image of North-Western Europe as a region of ecological modernization versus the view of the United States as a more traditional market-liberal region where green product and service claims are more often seen as a threat to fair market competition. We can also see this separation in the area of consumer regulation, movements and activism, where there have been far more protests and consumer boycott campaigns in US history, particularly surrounding the civil rights movement (Vogel, 2001, 2004; Micheletti, 2003). This pattern also emerges in more recent campaigns. Oppositional movements protesting against mass and mainstream consumerism appear much more visible and louder in the American marketplace. Examples of such protests include anti-consumerism, anti-television activism, and anti-advertising campaigns (see Cohen et al., 2005).

Although such distinctions are useful in several respects, the placement of regions on one of the two poles runs the risk of obscuring at least two things. Firstly, it partly ignores the oscillations between conflicts and agreements that take place in each region *throughout entire processes of policymaking*. Secondly, it does not take into account the variations between conflict and consensus that take place *across sectors* within the same country or region. Consequently, the above-mentioned aspects could take place also within a specific political sector at a specific level.

As the concept of political culture is broad, we try to delimit the notion of political culture as a latent 'readiness' among actors within a political setting to do certain things, for example, a 'readiness to negotiate' on policy issues among a broad group of interest groups. Within a policy context (country or sector), policy participants may learn, internalize, and be socialized into a certain way of making politics and policy, thereby developing an inclination to follow a particular tradition (such as using dialogue and negotiation rather than confrontation).

### Readiness to negotiate

With the phrase 'readiness to negotiate', we refer to a political culture in which actors that represent different interests have latent capacity and willingness to communicate and negotiate with each other. This readiness includes expectations that actors can affect each other by way of argumentation and that it is possible to reach compromises that are reasonable in the eyes of all involved.

Sweden has often been portrayed as a country with a 'strong state' and corporatism. Corporatist policymaking occurs in an organizational setting in which the state and well-organized and centralized interest organizations (typically labour unions and trade associations) assume a strong policymaking role and negotiate about welfare policies. Within such settings, Sweden has developed a consensus-oriented political culture with a preference for pragmatic problem-solving and reformist orientation (cf. Micheletti, 1995; Lundqvist, 1996). Scholars claim that a strong, responsive and selectively open state (which is open towards large interest organizations), as seen in Sweden, tends to absorb new ideas and private initiatives (cf. Kitschelt, 1986; Jamison et al., 1990). An example is the strong role of the state agency called Swedish Consumer Agency, whereas voluntary consumer associations are extremely small and insignificant in Sweden. As a comparison, consumer policy and information is an area in which we see a much stronger role for autonomous consumer associations in, for example, the United Kingdom and the United States. Labelling processes are market-based and normally gain from including many types of actors, such as consumer associations and environmental movements, which are not used to being involved in traditional state-led policymaking. A political culture marked by state interventionism and corporatism could therefore constitute an obstacle to private actors in developing alternative regulation. Although the last 15 years have seen a formal abandoning of corporatism (Lindvert, 2006), we should not expect big interest groups suddenly to change the ways of making policies, for instance by way of dialogue and negotiations.

However, earlier studies on 'joint environmental policymaking' (JEP) indeed indicate that corporatist traditions facilitate rather than hamper policy processes comparable to labelling.[50] Mol et al. (2000) analysed such policymaking in the Netherlands, Denmark, and Austria and argued that JEP is particularly well developed in countries with a political culture characterized by cooperation, moderately open policy networks, and consensus-building between private and state actors (Liefferink et al., 2000). Accordingly, a country with a tradition of negotiations and consensus could facilitate both policymaking such as environmental agreements as seen in the Netherlands (ibid.), and green labelling as seen in Denmark and Sweden.

A strong readiness to negotiate has been apparent in our Swedish cases of forest certification, seafood labelling, and organic labelling (whereas it has been less striking in the Swedish eco-standardization of paper and electricity). Moreover, the contrast with the American cases on forest certification and organic labelling is very apparent.

For example, the Swedish FSC process relates to a consensus-oriented policymaking culture which dominates in the sector, and which includes the state administration, the forest industry, and labour unions. Environmental protests regarding forestry escalated in the 1980s, but a turning point preceding the FSC process took place when a parliamentary committee was set up in 1990, in which EMOs could take part. Informants maintain that this policymaking process was important in that it stimulated parties to begin talking and listening to each other's arguments. This policymaking process paved the way for the FSC process in several respects (see Boström, 2003b); FSC actually re-invented a tradition characterized by dialogue, pragmatic problem-solving and negotiation among big interest groups in this sector. One informant from a labour union mentioned how surprised he was that it was possible to extend the Swedish spirit of dialogue and consensus to new groups within the Swedish environmental movement.

Our Swedish seafood case was selected as a critical case partly so that we could investigate whether this specifically Swedish political culture could play a role despite strong initial antagonism and mutual mistrust. Before seafood labelling was initiated in Sweden, there was a deep conflict between groups representing the fishing industry, on the one hand, and EMOs as well as groups representing sport fishing, on the other (Boström, 2006a). The Swedish fishing debate was impregnated with controversies. Initially, people from these different interest groups could not even sit in the same room. In the end, they nonetheless succeeded in agreeing on a compromise. What is interesting to note is precisely this readiness to negotiate. Through skilful organizing and mediating by the labelling organization KRAV (see also subsequent chapters for more analysis on this case), people learned – after a while – that it was possible to sit in the same room with opponents and to discuss contrasting viewpoints. We argue that the Swedish political culture of dialogue and negotiation is facilitating. Despite antagonism, groups simply know that it may become possible to communicate with opponents.

In the United States, the larger and more heterogeneous political-administrative system, together with deeper religious, ethnic, and socioeconomic gaps compared with Scandinavian countries, leads to a more polarized political culture (e.g., Christensen & Peters, 1999). The handling of deep social cleavages is often 'not that of compromise and bargaining but rather that of confrontation' (ibid., p. 136). Hence, there is in general less preparedness for dialogue and compromise among a broad group of stakeholders in this political context. In our comparative cases of GM, organic food, and forestry we see a significantly more polarized atmosphere in the United States. Cashore et al. (2004) report that the US forestry case shows the most polarized climate in all of their case studies (United States, Sweden, Germany, United Kingdom, and British Columbia in Canada).

An adversarial policy discourse climate such as in the United States is likely to foment strong polarizations and crossovers, as well as unproductive framings such as 'scientific truth' versus 'ideological superstition'. Such polarization generally prevents fruitful discussion among the many shades of grey in labelling, which are more easily identified within a consensual setting. On the other hand, within the Swedish consensual policy climate, discussions follow an eco-pragmatic metaframing (see Chapter 8), which in turn effectively excludes critical environmental themes by definition, without bringing them up for larger public debate. In contrast, a benefit of the confrontational political culture is that it facilitates a multi-frame debate (cf. Dryzek, 1993). A number of critical issues that are open for broad public and stakeholder debate in the United States are more or less absent in the Swedish discussions. These include epistemological and methodological issues concerning labelling practices, the survival of small-scale farming, ethical consumerism 'beyond organic', as well as the intriguing elements in the Big Three debate (we elaborate on these examples in subsequent chapters).

In sum, a specific political culture of pragmatism, consensus, and openness enables *latent preparedness for interest groups to engage in dialogue and make compromises, jointly looking for solutions.* Large interest groups in different sectors are inclined to search for novel solutions and compromises, even when actors may initially be strong antagonists. Experiences, awareness, and knowledge about the possibility of communicating are parts of a collective memory among big interest groups, including the environmental movement. Yet, it is far from evident that such a political culture is always helpful for enriching frame reflection and broad public debate.

### Readiness to support

Another 'readiness' variable that is relevant to our discussion is the readiness among existing political and regulatory authorities to accept or assume a strong role for private actors (or 'private authorities', cf. Cutler et al., 1999; Hall & Biersteker, 2002), including non-state labelling initiators. How can state and political actors support green labelling? What is their willingness to support such market-based tools? Is it important whether or not they express support?

State actors can provide both practical and symbolic support. Practical help may involve funding of green labelling, either the standard-setting process itself (e.g., in the Swedish seafood labelling case) or the actual certification (which has been done, e.g., in order to convert to organic food production; see Michelsen, 2001). State agencies may buy green-labelled products. They may participate as knowledgeable actors in standards development because they may have expertise on existing rules and on ecological conditions.

In addition to practical support, symbolic support, such as explicit approval or authorization, could be highly influential. To give practical support is an indirect way of giving symbolic support, which entails legitimizing green labelling indirectly. Yet, state and political actors may explicitly support non-state-driven green labelling. For example, organic production in Sweden has enjoyed support in various policies and statements, such as the government's statement that 20 per cent of Swedish arable land should be certified organic by 2010.[51]

Granting such legitimacy is not left unquestioned. The political culture of readiness to support a non-state-driven rule-setting process can vary significantly between countries and sectors. A tradition of broad public–private collaboration can be an important factor behind the likelihood of getting support for a private rule-setting initiative. Public authorities may interpret labelling initiatives as not respecting the authorities and the existing regulatory framework; the authorities may express concern about not being invited to participate; and government, authorities and industry players may agree that certification and labelling are superfluous (cf. Elliot, 1999; Constance & Bonanno, 2000; Elad, 2001; Cashore et al., 2004; cf. Chapter 6). In the United States, for example, the USDA and the government have a much more reserved attitude to organic food products and processes. The organic industry and the organic movement in the United States do not perceive any strong support from the government and its authorities, and, instead of qualifying it as an 'ecolabel', the US government frames organic labelling in the United States as simply 'a label based on consumer preferences' (e.g., Klintman & Boström, 2004).

When and why do state actors and political actors support green labelling? First, a specific political culture of broad joint policymaking may contribute to a supporting attitude, as we analysed above. Second, state and political actors may support labelling if such rule-setting harmonizes with broad political goals and policies. Therefore, it can be important that labelling initiators frame their initiative as a way to supplement state regulation, not to replace it. Likewise, an increasing focus on consumer issues in politics and policymaking adds to this tendency (cf. Chapter 2). Third, a tradition of state-centred policymaking can be an obstacle to a private rule-setting initiative. When comparing the Swedish forestry and fishery administration, the latter sector appears centralized, whereas the former sector has long since developed an administration with delegated authority, and with a preference for soft regulation. Given these varying traditions, it is no coincidence that we found stronger support for labelling among state authorities in the forest case. Fourth, if labelling processes are seen to be inclusive enough – or inclusive in the correct way – state actors may support the processes (Rametsteiner, 2002; Boström, 2003b, 2006a, b). State actors do often have an obligation to be responsive to a broad spectrum of interest groups. If non-state-driven policymaking and rule-setting are seen to be inclusive, state actors may recognize that many different interest groups including themselves have access. State actors may not require membership in the labelling organization, but may want to have clear insight and access, thus being able to share their views and concerns. For example, when Swedish state actors rejected the MSC, the state officials addressed what they perceived to be the closed and non-transparent nature of MSC.

Is it important that state actors express support for green labelling? Even if a non-state market-based governance scheme does not rely on states, states are powerful actors that may potentially challenge the legitimacy of a scheme (cf. Bernstein & Cashore, 2004). Again, the Swedish seafood labelling case shows the critical need for state support. Interviewees from WWF said there was no point in trying to convince state authorities or others to set up a Swedish working group following the MSC framework, because of their strong criticism. Their initiative failed partly because of a lack of state support, and it would take some years before new initiatives for a seafood labelling process were taken, this time within another organizational setting (KRAV) in which state actors have better access.

Because state authority is rendered legitimate through the principles of representative democracy, state actors' framing of private initiatives as welcome, relevant and important can help to create the public perception that a private initiative is more 'public', 'democratic', 'trustful', and 'accountable' (Hofer, 2000; Rametsteiner, 2002; Boström, 2003b, 2006a, b; Wälti et al., 2004). It is not evident, however, that strong state support really establishes more legitimacy for the non-state-driven labelling project. A degree of autonomy and distance is at the same time often an important basis for legitimacy. If a rule-setting process is run outside state arenas, certain actors may assume that it is safer to participate and share information. An aura of voluntarism and autonomy surrounds the project, and actors do not completely commit themselves to going the whole hog. EMOs often criticize state-centred policymaking for being inert, and they therefore want to keep state authorities at a distance. The challenge is to find a balance, because drastic exclusion of state actors can reduce legitimacy. The non-state-driven initiative needs practical and symbolic support, at the same time as it may have to be run independently of state intentions and directives.

## Readiness to regulate

If political actors and state actors do not embody a political culture of readiness to negotiate or a readiness to express support for private regulatory innovators, they may nevertheless be able to support labelling through *a readiness to regulate labelling*. For instance, the restrictive American attitude towards regulation is, indeed, mixed with a certain political culture of readiness to regulate and standardize, as a response to protest campaigns and 'public input' (Dryzek et al., 2003). David Vogel (2001) maintains that the United States, before the mid-1980s, did to a great extent lead the development of consumer- and environmentally related regulation, frequently with reference to a precautionary and risk-averse approach. The more the American public has tended to worry about a particular risk, the more strictly American policymakers have been likely to regulate it. For example, during the 1970s, 'US agencies designated as carcinogens a number of chemicals that most European officials did not consider a cancer risk to humans' (Vogel, 2001, p. 2). They banned pesticides, dioxins and food additives to a larger extent than did European counterparts. Furthermore, US regulation pioneered in environmental disclosure and right-to-know legislation through the establishment of Toxics Release Inventory in 1987 (Van den Burg & Mol, 2008). This took place well before the signing of the Aarhus convention in 1998, which required members to establish publicly accessible databases with information on the environment.

It is also a well-known phenomenon that 'free' trade and commerce require rules of the game. Thus, a political culture praising market liberalism and individualism does not automatically stand in opposition to regulation or legalism (Christensen & Peters, 1999; Egan, 2001). Indeed, the United States is often viewed as the prototype of the regulatory state (Majone, 1996; Egan, 2001).

Since the mid-1980s, however, the political salience of consumer-related and environmental regulation has gradually increased in Europe whereas it has decreased in the United States. The United States is considered to have been bypassed in this respect by the EU and European countries, by establishing a range of consumer and environmental standards that are stricter than their American counterparts (Vogel, 2001). Compared with the EU and several of its member countries, mainstream policymakers in the United States, at the federal level,[52] have so far shown very little interest in promoting either sustainable development or sustainable consumption (Cohen et al., 2005). The EU clearly embodies readiness to regulate in the areas of agricultural, consumer, and environmental regulation. Indeed, as a political organization the EU is more similar to the US federal structure than most single European countries are. As Majone (1996) notes, the EU is primarily a regulatory state, which implies that issuing rules is its most important means for shaping public policy (see also Egan, 2001). Vogel argues that earlier regulatory failures, associated for instance with mad cow disease, and the subsequent lack of public confidence in the European food supply, to a great extent explain this change in preparedness for regulation (see also Carson, 2004). The most visible example of the new regulatory mode, according to Vogel, is the regulation of genetically modified crops. He claims '[i]t is impossible to exaggerate the significance of the regulatory failure associated with BSE on the attitude of the European public toward GM foods' (Vogel, 2001, p. 12). The strict European GM labelling, which stands in sharp contrast to the negative American attitude, is an example of that. Similarly, the American faith in the capacity of risk assessments for categorising a product or technology as 'safe' or 'unsafe' significantly contrasts with the situation in Europe, where both citizens and officials to a greater extent embrace the precautionary principle (see quotation of Vogel in pages 167–8).

Still, it is important to stress that this 'negative American attitude' towards regulating genetically modified crops is specifically a negative attitude of the US Government (at least the FDA and the USDA) and the GM food industry, and by no means the attitude in the United States as a whole. In this case, the adversarial US policy climate is reflected in the non-GM movement and organic food movement actively struggling against GMs. As Roff (2007) points out, at least 60 groups are actively involved in this agbiotech opposition. These groups range from single-issue organizations such as the Genetic Engineering Action Network to international multi-issue organizations such as Friends of the Earth and Greenpeace.[53] From a regulatory perspective, however, Roff notes that these anti-GM efforts have a strong market focus. Although the groups typically target the government and the regulators, it is much more common to call for increased monetary political consumerism (to persuade consumers to buy organic foods). Furthermore, these non-GM groups frequently use the tactic of asking consumers to write letters to food chains and producers protesting against their GM use. In effect, this market focus among protest movements in the United States is in line with a lower readiness to regulate GMs in the United States than in Europe; the non-GM movement in the United States seems to assume that the public has a higher readiness to avoid GMs in the stores though choosing organic than the Government has readiness to *regulate* GMs.

Readiness to regulate is a political culture variable; not to be confused with existing rules and regulations, which are one of our four general context factors. Of course, readiness to regulate may lead to the implementation of new rules and regulations. However, existing rules and regulations need to be analysed as such, including the way they shape and confront labellers. This is what we set out to do next.

## Existing rules and regulations: enabling, protecting or suppressing?

The second general context factor that can both facilitate and constrain new rule-making initiatives is existing transnational, national, and/or sector-specific rules and regulations. In the global context there is an abundance of transnational rules and policies aimed at a variety of organizations (Djelic & Sahlin-Andersson, 2006; Boström & Garsten, 2008). Yet, most of them are formulated in very general and vague ways, giving great scope for interpretative flexibility. Green labelling has a double relation to the emergence of abstract transnational and national rules. On the one hand, this vagueness and ambiguity of existing regulation is a fertile context for EMOs and other initiators for various framing efforts. Labelling initiators identify and frame many shortcomings and failures of existing regulations, which function as a motivation for alternative regulatory initiatives including labelling. On the other hand, existing soft regulation (Mörth, 2004) may provide indirect legitimacy to labelling. At the same time as labelling initiators frame the insufficiency, vagueness, slowness, softness, and ambiguity of existing transnational and national regulations and policymaking, they also pick arguments and relate their own standards to broad policies and regulations.[54] Hence, such transnational actors as the UN, ILO, ISO, and WTO are simultaneously important targets of criticism and sources of legitimacy.

Sometimes, labellers more or less have to relate to, and approach, consistency with existing regulatory frameworks (cf. Streeck & Schmitter, 1985; van Tatenhove et al., 2000; Rametsteiner, 2002; Bernstein & Cashore, 2004). Certain ('hard') rules and regulations simply compel labelling initiators to take into account various things; otherwise labelling may turn out to be illegal. Other rules are voluntary in the formal sense ('soft regulation'), but they may have received a strong legitimate status as morally binding. In addition, labelling relates to existing rules and regulations in that the initiators usually hope to establish a scheme beyond the minimum-law level in order to visualize good environmental performance – best practice within a sector. The symbolic differentiation between 'green' and 'conventional' products is then supplemented by a kind of symbolic differentiation between 'green' and 'conventional' rules and standards.

### The example of free trade

One of the challenging issues for green labelling in general is how this kind of instrument affects free trade. Among certain WTO members, especially from developing countries, there is a fear that labelling schemes constitute disguised protectionism, which, for example, has been intensely debated in relation to the labelling of seafood and forest products (Deere, 1999; Bernstein & Cashore, 2004; Gulbrandsen, 2005b; Oosterver, 2005). In principle, WTO only accepts regulation of global trade if it concerns characteristics of the product itself. Rules must not be based on distinctions between products in terms of their production or processing methods (PPM) if the physical characteristics of the *products* themselves are equal.[55] Green labels are typically based on PPM criteria. From a strict interpretation, WTO rules firmly delimit what products can be subject to green labelling. However, *voluntary* labelling schemes are largely beyond the direct jurisdiction of the WTO, and this type of instrument does not deny market access in principle (Deere, 1999; Klintman, 2002a; Bernstein & Cashore, 2004). Moreover, whereas they are built on PPM criteria, labelling programmes such as the MSC and FSC could be seen as concrete responses to the dominant WTO discourse because of their global application, which enables global flows of products (Oosterver, 2005).

Labelling is generally considered to take place outside the scope of the WTO, as long as governments remain uninvolved (Bernstein & Cashore, 2004). A complicating factor is that governments actually are often involved in one way or another, as we have shown in this chapter. For example, many governments are forest owners. They may consider forest certification and labelling in their procurement policies. Therefore, the trade discourse in the WTO can very likely have an impact on, for example, the EU regulation and national regulations, which in turn may affect labelling or green procurement policies (cf. Meidinger, 1999; Gulbrandsen, 2005b; Boström, 2006a). For example, several labelling programmes, including organic labelling, face difficulties in introducing a limit on transport distance, which would conflict with WTO and EU free trade rules. Such information could be easier to include within other eco-standards, such as environmental declarations, when the intention is only to inform consumers (cf. Nilsson, 2005). From all these pressures it follows that green labellers have to consider how a local scheme affects trade issues. Labellers feel pressure to develop methods and principles (such as mutual recognition, international harmonization) that enable transnational trade. This pressure is especially serious for local or regional schemes,[56] as the Swedish seafood labelling case illustrates.

---

*Example of free trade concerns in the Swedish seafood labelling case*

The very 'Swedishness' of the Swedish seafood labelling project appeared as a main source of disputes (see Boström, 2004b, 2006a). For example, the pressure from the pro-free-trade commentators resulted in quite a major shift in how to formulate standard criteria for technology. Different catch techniques have been a major topic in the Swedish debate about fishery (e.g., Johansson, 2003). Several debaters perceive some methods as environmentally bad, almost by definition. They metaphorically associate techniques such as trawling with fishermen vacuum-cleaning the sea. Accordingly, social movement actors and debaters pressed to include strict standard criteria, such as forbidding trawling or drift nets. However, the standard states only that by-catches and the damage to the marine environment should be minimized. The project gradually approached a standpoint that no method or tool should be dismissed beforehand. The focus should be on the function of the tools rather than the tools as such. The same tool may be damaging in one marine environment, while it may be unproblematic in another. A certification body will assess case by case whether a certain technique for a certain stock can be approved, and a Council of Experts will provide the advice. This allows for flexibility in interpretation. It goes in a WTO-friendly direction, something fully in line with the intentions of the standard-setters. If the regulatory framework would disqualify, *a priori*, certain techniques, and thereby some sections of the fishery – which certainly was demanded by different stakeholders – the standard would be more vulnerable to criticism contending that the standards constitute technical trade barriers.

---

## Preventing the misuse of terms or suppressing autonomous definitions?

State actors may set up rules in order to prevent misuse of terms, such as the labelling of conventional food as 'organic'. In both the United States and Europe, state authorities have seen the need to establish rules that provide a protected and standardized definition of organic principles. An important rationale behind the regulation of organic farming in both the United States and the EU is to prevent inconsistent and inappropriate marketing of the term organic. Such regulation is intended to make consumers trust that food labelled as organic complies with certain requirements. It also protects organic producers from unfair competition.

An important difference between the European and US regulations, however, is that the European regulation allows private standard-setters to define additional criteria and set threshold levels above the European minimum level. In contrast, a state in the United States is not allowed to require higher organic standards than the USDA does federally, unless there are specific environmental conditions in a certain state that necessitate stricter state standards. These federal restrictions on organic improvements at the state level have led to strong reactions from the organic coalition. The federal rules indeed protect the label from misuse, but they simultaneously impinge upon the organic movement's ability to define autonomously what is organic, and suppress the continuous development of the label and its criteria (Boström & Klintman, 2006a).

However, organic advocates in the EU also worry about centralization of organic standardization. Informants from KRAV maintain that a great amount of their work since the early 1990s has been about adjusting their rules to the EU regulatory framework, which they regard as counter-productive and threatening to their autonomy. These tendencies intensified during the recent EC-led revision of the EU Organic Regulation 2092/91.

---

*Example of centralization tendencies in EU organic regulation*

One of several strongly criticized proposals was that the EU should include a mandatory labelling with an EU logo, which existing labelling programmes would have to use. Even more criticized was the suggestion to forbid labellers to state that separate labels such as KRAV's have stricter environmental requirements than the EU's minimum rules (such as in the US case). Another heavily criticized suggestion was that all certification bodies in the EU should be given the right to use each other's organic labels, which would entail that separate labelling organizations, such as KRAV, lost control over their own labels. At the European level, IFOAM's EU group struggled hard to counteract proposals that they thought would undermine the autonomy of the organic movement. Likewise, at the national level, in Sweden and elsewhere, a broad group of actors including KRAV, EMOs, associations for both conventional and ecological farmers, various agencies and even the government united in strong opposition. The coalition with IFOAM at the centre saw the EC's suggestions as attempts to confiscate private labels. They claimed that the EC lacks an understanding of the positive dynamics of separate labelling schemes. They also raised deep concern over the way the proposal had been developed, for instance regarding its suddenness and the poor consultation with relevant stakeholders, including the broad organic network. The coalition had some success in that it stopped some of the most serious suggestions, including those mentioned above, except the idea of introducing a common mandatory EU logo.

It is actually only in the agricultural/food area that we observe such intensive readiness to regulate, such deep involvement of state actors in protecting (or suppressing) the labels from misuse. Why are regulators particularly inclined to regulate organic food in Europe and the United States? To be sure, agriculture is a sector that traditionally involves a great deal of state interests, in both Europe and the United States (Moyer & Josling, 2002). There is also much cross-border trade in the sector. In the United States, one goal behind the standardization has been to facilitate interstate commerce in fresh and processed food that is organically produced. In the case of organic food, the increasing interest among American agricultural actors in moving closer to EU organic standards for economic and trade reasons has also been a strong motive for the standardization (Golan et al., 2000). In the EU, free trade *within* the EU's single market is also a strong rationale behind the EU regulation (Vogel, 2001; Moyer & Josling, 2002). Nevertheless, we see trade in many other areas too, for example forestry, in which we find no equivalent hard regulation. One additional important factor probably relates to previous regulatory failures in food issues (e.g., BSE) and the subsequent lack of public confidence in the European food supply, in turn leading to a special readiness for regulation in the food area (cf. Vogel, 2001).

In sum, regulators want to protect producers and consumers from misuse of terms along with these other trade-related objectives. However, such regulation may also contribute to an institutionalization of the labelling process to such an extent that it is relevant to talk of suppressing rather than protecting the labelling – or a kind of 'regulatory occupation from above' (cf. Boström & Klintman, 2006a). This institutionalization restricts the development of new labelling criteria and may impinge on the more general debate and reflections on the labelling and green consumerism. Looking from the positive side, such regulatory arrangements may positively trigger efforts to initiate new labels, such as 'beyond organic' labelling (Barham, 2002; Guthman, 2004), a tendency especially seen in the United States.

## Soft regulation, deregulation and green labelling

The Swedish forest certification case illustrates another, nearly opposite, way in which state rules can affect non-state-driven labelling. In Sweden, a new Forestry Act was implemented in 1994 with the innovation that environmental goals were given the same status as economic goals in forest policymaking and management (Boström, 2003b). However, the early 1990s were also marked by strong deregulation tendencies, and the new forest policy was consequently influenced by demands to use 'soft' policy instruments. Only a few detailed rules as to how the production goals and the environmental goals should be attained and balanced were included. The new forest policy implied freedom for the forest industry to develop its own methods for attaining the general goals. At the same time, the forest industry was expected to take responsibility and relevant measures. Otherwise, the state could step in later and introduce harder regulation, a possibility that governmental actors expressed. The new forest policy and legislation created a formal basis for the development of new policy instruments and methods in forestry, including forest certification and labelling. Informants representing forest business said that they felt a need to show that they really took responsibility. Otherwise, they could expect stricter legislation. Accordingly, being certified under the FSC would be seen as evidence that companies were indeed meeting both the environmental and production-oriented goals (see also Cashore et al., 2004).

In addition, it is important to emphasize that some deregulation in this sector was necessary for introducing the forest certification alternative. Prior to the revised Forestry Act, the strict emphasis on mandatory production goals in Swedish forest policy (which included, e.g., strict obligations to thin out and clear forests) made it impossible, indeed illegal, to introduce alternative (environmentally friendly) forestry methods.

The forest case shows that a 'soft regulation' approach that explicitly upgrades environmental measures can facilitate a labelling process. Deregulation can facilitate labelling in that it paves the way and sets a formal ground for the initiation of alternative rules, for re-regulation, such as the initiation of voluntary labelling standards. Furthermore, the threat of 'hard regulation' has never been absent. Many of the strengths of this strategy connect with the fact that forest enterprises felt their responsibilities to do something 'voluntarily' for the environment.

Likewise, green electricity labelling is largely dependent on a deregulation in the electricity sector in several countries, including Germany, the Netherlands, and Sweden. In addition to the argument that deregulation entails better competition and thus lower prices for consumers (something that has not yet been confirmed in Sweden), the inclination of conscious consumers to choose a green electricity label, rather than nuclear power, was a major argument among those who wanted a deregulated electricity market.

> A key to the outphasing of nuclear power is the deregulation of the electricity market ... With the constantly growing environmental concerns among the public ... the consumers who think green may rearrange the energy market faster than we may imagine. (Per Ribbing in *Miljöeko* 2002/5, our translation)

Despite these – sometimes prematurely green – embraces of electricity deregulation, cross-national studies emphasize that substantial development of renewable energy still requires policymaking, not least a sense of pressure that policymakers might start to partly 'regulate' in new ways or to place extra tax on non-renewable energy sources, to stimulate renewable energy generation. In addition to the incentive of creating a green image for the company, this sense of regulatory uncertainty is – at least for some private companies – a factor motivating them to choose electricity with a green label, as studies in Germany, the Netherlands and Sweden indicate (Gan et al., 2007). Public policy measures, the regulatory environment, green taxes, external costs, and future subsidies are examples of policy-related pressures that should ideally take place in collaboration with governments at local and national levels, and with an active role for NGOs. In the deregulated electricity market in the United States, by comparison, researchers did not see the same policy-related pressure on companies:[57]

> Without significant increases in fossil fuel prices, much more stringent environmental regulations, or significant changes in electricity customer preferences, green electricity markets are likely to develop slowly in the United States. (Gan et al., 2007, p. 144)

## Creating labelling metarules

The analysis above should have illustrated that labelling does not occur in a regulatory vacuum. Labellers themselves quickly notice that. One response to the challenges of existing rules and regulations, including existing powerful transnational regulatory organizations, is the establishment of meta-organizations[58] for labelling organizations. Such meta-organizations (i.e., organizations with labelling organizations as members) issue rules on how to conduct labelling.

For example, the Global Ecolabelling Network (GEN) was established in 1994 to coordinate, improve, and promote national green labelling globally.[59] Another example is ISEAL Alliance, which was founded in 1999 by eight organizations of, as it claims, leading international standard-setting certification and accreditation organizations that focus on social and environmental issues; among its founding members were FSC, MSC, and IFOAM.[60] Here too the strategies are assistance, engaging in debate, and facilitating exchange and harmonization among members. ISEAL has recently developed its Code of Good Practice for Setting Social and Environmental Standards, which is mandatory for ISEAL's full members, and which emphasizes such aspects as openness, transparency, and broad stakeholder dialogue in standard settings.

According to its Executive Director, Sasha Courville, an important role for ISEAL is to monitor what is going on among other global standard-setters such as WTO, OECD, and ISO (interview). ISEAL was established because the individual members found they faced similar problems and challenges. They could share experiences and try to identify best standard-setting and certification practices, and they all committed to high standards of credibility. A challenge for ISEAL is the rapid growth of certification and labelling in general. A risk confronting all serious certification and labelling organizations is therefore that any flaw or scandal among any labelling organization could 'harm the entire movement'. So ISEAL members want to communicate that they all have highly credible systems. They also share a concern about how, for instance, the WTO's free trade rules and other global rules and doctrines could affect labelling. Consequently, ISEAL engages in both looking at the relations among different labelling systems and the relation between labelling and other eco-standards and regulations worldwide. It is clear that ISEAL has been established on the transnational scene to carve out some regulatory space for 'credible' certification and labelling instruments, in a partly conflicting relationship with other global players such as WTO and ISO. ISEAL does not try to attack or compete with these organizations, rather to find ways to relate to their regulatory frameworks without compromising too much their goal of rewarding high social and environmental standards.

## Organizational landscape

The third general context factor is the organizational landscape. We argue that the constellation of actors and their existing power relations and resources within a country and sector are important for understanding labelling processes. How power resources (material, cognitive, and symbolic) and relations of autonomy and dependence (Ahrne, 1994) are distributed within the organizational landscape can affect whether and how labelling is implemented, and it can shape the debate. In Chapters 9 and 10, we will analyse organizing processes at greater length. Here we just briefly introduce how such processes link to existing organizational structures.

A first sub-factor is the strength of the civil society, which plays a key role in green labelling. The concept of civil society includes organized interests such as NGOs or SMOs. Such organizations are comparatively strong in the organizational landscape in both the United States and Northern Europe (e.g. Micheletti, 1995; Lundström & Wijkström, 1997; Zald & McCarthy, 1997; Mertig et al., 2001; Ott, 2001; Dryzek et al., 2003). Too much emphasis on the pluralist/adversarial (United States) versus corporatist/consensual (Northern Europe) distinction may miss this important point. A relatively 'mature' national organizational landscape, with a strong civil society, especially in the environmental field, is likely to favour labelling initiatives; such a landscape includes resourceful EMOs with substantial symbolic, cognitive, and organizational capacities, which they can use to initiate and participate in labelling organizations and to stimulate green political consumerism. In Chapter 9 we discuss how such organizations may play constructive roles as outsiders, advisors, and decision-makers within and around labelling arrangements.

Regarding the significance of an organized civil society in the environmental field, we find a striking contrast between the similar neighbouring countries Sweden and Norway, a contrast that largely explains the different attitudes to green labelling. Although the organized civil society is strong in both these Scandinavian countries, by global standards, the environmental movement is much stronger in Sweden in terms of membership and public policy impact.[61] Gulbrandsen (2005a) sees the strong, progressive, and agenda-setting role of Swedish EMOs as one factor explaining the much greater support for FSC certification in Sweden than in Norway (where a competing, industry-led programme under the PEFC standard was initiated). Norwegian EMOs adopted the strategy of their Swedish counterparts but soon found that the forest owners' association developed their own programme, and simply rejected the EMOs' proposals. Such differences could also be explained with reference to trust. The lower extent of political consumerism in Norway than in other Northern European countries has been explained – in connection with food products – by a comparatively high public trust in food experts, retailing systems, and traditional political institutions (Terragni & Kjærnes, 2005). If people place very high trust in traditional institutions there will probably be less demand and willingness to look for alternatives.

When analysing the impact of organizational landscape, we should not merely look at civil society. The way industry sectors are organized may also have an impact on green labelling, in various respects. The organizational landscape within a particular sector may entail few and big companies, as can be seen in the forest and paper case, or many small companies, such as in organic farming. By looking at our cases, it is apparent that 'successful' labelling can occur in quite varied types of industry structures. Accordingly, the character of the organizational landscape does not determine whether or not labelling can be established.

However, labelling initiators may benefit from developing an understanding of the specific structure, including the power relations, within a sector. An organizational landscape with few, well-represented and unified companies can create both opportunities and problems. It centralizes power resources, and the critical issue for labelling initiators is the capacity to mobilize precisely these power resources. Otherwise, labelling initiators may be effectively counteracted. We will analyse this further in Chapter 9 by comparing the US and Swedish forest certification.

An organizational landscape with few, well-organized companies may also affect how symbolic differentiation is played out. As we have maintained, the development of a credible labelling scheme requires that it is possible to differentiate among business actors. In other words, it must be possible to distinguish between those that can be subject to certification and those that cannot; and such a symbolic differentiation can be difficult in a monolithic structure. An industry/market structure in which only a few competing sellers operate, and in which they have developed exclusive cooperation (without necessarily being illegal cartels), can be extra problematic. For example, the Swedish pulp and paper sector, which is characterized by oligopoly, was a structural condition behind the companies' joint abandonment of the Swan label, and their introduction of their alternative Paper Profile. Informants from the Swan saw this as an industry boycott, a kind of cartel (Nilsson, 2005). The companies are used to acting concordantly; they therefore had good capacities to make agreements on developing a common eco-standard.

An interesting aspect in the 'negative' paper case is also that retailers in the middle of the production chain were weak in relation to a few strong producers. There were a few wholesalers who criticized the environmental declaration (Paper Profile) and instead expressed support for labelling via the Swan. However, the paper producers owned these wholesalers. The wholesalers hence had a subordinate power position and could not convince their parent companies to change their negative attitude towards green labelling. In Chapter 9, we will emphasize the important role of proactive business actors in the middle of the production chain, particularly retailers, as proponents of labelling. Consequently, the structure of the retailing system can also affect implementation, development of labelling criteria, debates and symbolic differentiation.

When comparing organic production and retailing in the United States and Sweden, one can find certain structural differences. As in a number of other European countries – Denmark, the United Kingdom, Finland, Austria, Switzerland (Torjusen et al., 2004) – Swedish organic products have been integrated into the conventional retailing system, in which most people buy most of their food. This integration helps to normalize organic consumption as such, and it is a critical factor for increasing both supply and demand.

In the United States, this process of convergence has been slower, and the organic and 'conventional' sectors have [until recently] appeared to be more differentiated. For a long time, to a larger degree than in Northern Europe, American retail sales of organic food used to take place in stores promoted as 'natural products stores' or 'organic super-markets' (Vaupel, 1997). In Sweden, the traditionally more unified food market structure has enabled efficient distribution of labelled products to a majority of shops all around that country. At least until a few years ago, such structural differences could help create a different character of the labelling debates in the two countries; this would be one – among several – structural conditions accounting for the more polarized debate in the United States.[62] Still, it will be interesting to see whether the changing organic market structure will make the US debate less polarized (Boström & Klintman, 2006a).

'Normalization' may be crucial for the growth of an organic market. However, normalization may occur at the expense of visible alterna-tives. Consumer emphases on local food (e.g., keeping the 'food miles' low), quality aspects related to 'handicraft' small-scale processing (e.g., sparse packaging), or personal trust relations, might be harder to meet inside a standardized food system, according to Torjusen et al. (2004). They observed the more prominent role of specialized food stores in the organic sector in other European countries than those listed above – for example, Germany, the Netherlands, Belgium, Italy, and Greece. Such retailing systems may promote particular ideologies and certain values (and such aspects mentioned above) as they link to particular cultural or political networks or are based on a cooperative structure.

## Materiality and technology

When we presented our four context factors concerning green labelling at a seminar, all our helpful readers were content with the first three. Yet, they were sceptical of the fourth one: materiality and technology. One of our colleagues came up with the question: why would labelling schemes be different *merely* because fish differ materially from forest and electricity? He urged us to beware not to claim that materiality and technology determine the policy outcomes. According to him, it would be an analytical fallacy to assign materiality and technology an independent, causal power for social and policy-related schemes such as labelling.

We agree with this call for caution. For several decades, scholars in the human sciences have been well aware of the dependency of perceptual filters – including sociocultural frames – on human perception (Goffman, 1974). Experience (of 'lay' persons as well as 'experts') is always situated and, indeed, mediated through cultural ideas and norms. Furthermore, since policymaking is always based on situated and mediated experiences, the material cannot directly determine policies. We also agree that the ways that policy actors, experts, EMOs, and the general public *frame* the materialities are significant for the way they develop and organize green consumer policies.

However, we find it important to take materialities and technology into consideration as external factors that may play a role in labelling, albeit not in a reductionist or deterministic way.[63] In what follows, we discuss size/technological complexity, mobility, and perceptibility.

First, it is relevant to return to the paper case when discussing *size* and *technological complexity*. Reliance on heavy investments (due to a dependence on heavy machinery) may appear to restrict flexibility to the extent that labelling is not seen as applicable in the industry sector. The pulp and paper sectors have industries with heavy investments. Companies that make big and expensive investments can face difficulties in complying with labelling expectations. As an example, the Nordic Swan makes a comprehensive revision of standard criteria every five years. Business actors in the paper industry claim they must have much longer time horizons than five years for their investments. This argument provided one of the key incentives for the industry to abandon the Nordic Swan and develop Paper Profile. In sectors that we have not yet studied – the medical, biotech, and chemical industries – we speculate that the argument of investment size and techno-scientific complexity will be used increasingly among industries that want to reduce the speed with which green labelling criteria are altered.

Second, the nature of the resource that is the object for labelling can have important implications for knowledge and debates surrounding it. In seafood labelling many actors – from fishermen to policymakers – regard the fish resource itself as a complicating factor. A basic difference between the fish resource and the other agrarian sectors, such as forestry and agriculture, is that the latter cases concern stationary resources. One informant from WWF Sweden who has been engaged in both the seafood labelling case and the forest certification case argues that this is one basic reason why the knowledge challenges and controversies have been so intense in the former case.

Third, perhaps an even more intricate matter is the perceptibility of the services, products and production processes subject to labelling. We have elsewhere examined the challenge of exposing unsensed risks in production processes that are detached from the everyday perception of green consumers (Klintman & Boström, 2006, 2008). If the products and services in themselves revealed their environmental, health-related and social implications, no labelling schemes would be needed. Still, there are indeed differences in the perceptibility of various products and production processes. People in our case studies have strongly emphasized how certain products and production processes in themselves may make policymaking and green consumer choices easier or more difficult. As to the end products, Darby and Karni (1973) have coined three terms of interest to the understanding of green products. *Search goods* can be fully evaluated by the consumer, by merely looking at the product in the store. *Experience goods* can also be fully evaluated by the consumer, but he or she needs to use the product to reveal its qualities. *Credence goods*, finally, are impossible for the consumer to evaluate even after using the product.

Researchers have made much use of Darby and Karni's distinction in consumer and marketing research through the years. Still, we argue that it needs to be nuanced and developed further in order to be useful for understanding green consumer policies. These authors erroneously imply that one may easily place most products under one of the three concepts. Organic food, for instance, could arguably be seen as any of the three types of goods. The researcher needs to examine the framing that operates in each case. It is typically the way an actor frames a product, instead of the physical nature of a product, that determines its categories.

Even though it is sometimes a contested issue, it is fair to say that producers and consumers treat certain organic food as search goods. When the consumer notices the organic label, some moderate and visible quality differences between products may strengthen the legitimacy of the label (Klintman, 2006). An organic tomato should be redder than a conventional tomato. However, as the eco-pragmatist pole of green consumer debates stresses (see next chapter), the differences should not be extreme. In addition to their 'natural' feel, organically labelled food products should also have a 'normal' feel to them; differences between labelled and conventional products should be barely visible. They should not be unusually pale, repugnant or in other ways fundamentally different from conventional products. Furthermore, labelled food should not be excessively colourful, since strong colour often signals artificial ingredients. It is also problematic for marketers to suggest that it is possible to taste differences due to 'organic', extra-healthy substances. Instead, the label ought to do the primary job of distinguishing ecological foods from conventional ones.

Organically labelled fruit may also taste better. Alternatively, we may perceive that it tastes better because we have been informed by a label that the fruit is organic and we may expect such fruit to taste better (Ekelund & Tjärnemo, 2004). Furthermore, in our empirical study, some consumer groups and EMOs maintain that organically produced foods are of higher quality than other foods (although 'quality' has several, sometimes contradictory, meanings); other actors argue against such claims. In green advertising, it is sometimes claimed that the labelled product also indirectly gives the conscious consumer a better sensation (Klintman & Boström, 2008). The very awareness among consumers that they purchase products causing less harm to the environment, farmers, and animals will pay off by making them healthier and happier. In this light, the advertisers do not want us to read the slogan on fair-trade-labelled coffee – 'with better aftertaste' – merely metaphorically.

That organically labelled food, through framing, can be seen as search, experience, and credence goods (although mainly the last category) may be a reason why food is a particularly hot topic in green labelling in general. The material truism that we eat food plays a key role here. It is a basis for extensive public worries over food crises, which in turn pave the way for a risk culture, and create demand for green labelling efforts.

Another modification of Darby and Karni's terms that is needed in order to apply it to green consumer policies concerns credence goods. Whereas many – perhaps most – products subject to green labelling are credence goods, we need to make a further distinction. On the one hand, there are 'measurable credence goods', with experts claiming to be able to perceive material differences in products, for instance certain organic foods, textiles, and the energy efficiency of products. The other type, 'absolute credence goods', needs a few words here. Among such goods, no material difference can be measured between products. Absolute credence goods include green electricity, forest products, and paper (the latter when it comes to the electricity efficiency of paper production). In electricity and forestry, the final products are undoubtedly identical regardless of whether they are labelled or not. The 'green' electricity consumer gets the very same electricity as consumers who have not chosen electricity with a green label. Interviews with consumers have indicated that this lack of visibility through concrete product separation relates to people's mistrust (Klintman, 2000; Klintman et al., 2003):

> I: Some talk about green electricity. Is that something that you have thought about?
>
> IP: Yes, and it is an incredibly stupid idea!
>
> I: How do you mean?
>
> IP: I can't understand how people can pay extra when all ends as the same 'soup' [with the same electrons coming through the wall]. (Klintman et al., 2003, p. 67)

If the difference across end products is not material, but mainly administrative, many actors in our studies conceive the label as insufficient for visualizing and motivating broad, consumer groups to make the environmentally sound choice.[64]

As a final note, we should mention that, despite the material and technological challenges that are part of certain sectors more than others, we hold that it is possible to resolve most challenges in terms of policy adaptation, organization, and frame-reflective deliberation on labelling schemes, along with information and communication adapted to various consumer groups.

## Conclusion

In this chapter, we have illustrated how various context elements matter while not determining labelling outcomes (implementation, debates, symbolic differentiations, etc.). Therefore, we do not want to conclude by insisting that a specific type of context determines successful labelling. Similar kinds of labelling take place in different contexts and different kinds of labelling can take place in one context. There is no *one* best way that leads towards successful labelling. Perhaps the most important implication of our findings – developed through systematic comparisons – is that green labelling actually works in very different policy contexts. A policy implication of our analysis is that labelling initiators should be aware of the specific patterns in the country and sector in which the initiative is taken. In an increasingly globalized world, with multilevel governance structures, we still need to be aware of how local arrangements are influenced by existing political cultures, regulations, and organizational landscapes, as well as by materiality and technology.

The American standardization of labels is geared towards the creation of antagonist coalitions and public mobilization. As green labelling is a market-based and consumer-oriented instrument, we might have expected successful and smooth implementation in such a market-liberal country. In cases of organic food and GM food, the US federation – in a vein of state monopoly – has restricted labelling in various ways, which in effect constitutes a restriction on the space for green consumerism. Likewise, the introduction of FSC in the United States was met with muscular counteraction from a unified forest industry. An interesting exception, however, is the case of green mutual funds, in which the United States presents a front-running case. SRI funds in the United States have been developed in an organizational landscape and with existing rules that favour product differentiation of portfolios for various groups of investors. In general, the banks and other investor companies in the United States were early in offering a broad variety of mutual funds to investors. Moreover, one may speculate that the strong tradition in voluntary activities, such as charity giving and philanthropy, constitutes a political culture facilitating the development of SRI funds. Nevertheless, we prognosticate that the strong existing rules and regulations aimed at ensuring a fair competitive market will be the basis for closer scrutiny of the SRI fund market, similar to governmental scrutiny of (other) green and ethical labelling claims.

Sweden and Northern Europe, on the other hand, provide a state-centred policy context, which could have constituted an obstacle to private actors in developing market-based policy approaches. Yet, the Swedish policy context, with a political culture fostering a 'readiness to negotiate', friendly and collaborating relations between public and private authorities, and a relatively strong role for well-organized civil society actors, appears fruitful for green labelling implementation, market impact, and debate among large interest groups. However, it is less evident that this policy context facilitates broad public debate and reflection on green political consumerism in general.

Drawing upon the figure at the beginning of this chapter, differences of organization and framing, which we analyse in subsequent chapters, may be linked to differences in policy context. In organizational terms, polarized political culture may entail a stronger 'fighting spirit' of various groups through, for instance, protest mobilization. This is closely tied to democratically invigorating frame-critical debates that might arise in the more polarized policy context, of which we have seen examples in US cases. Still, this productive development presupposes that open processes of mutual learning take place, rather than constant repetition of fixed or 'crossed-over' arguments across poles (Klintman, 2002b; Klintman & Boström, 2004).

Similarities across sectors in the same country show that national contexts do indeed matter (e.g., a general readiness to negotiate in Sweden and a generally more polarized debate in the United States), although we should not exaggerate this point. There is a clear difference across sectors in the same national context as well. Furthermore, each sector is entangled with multilevel governance. Organic labelling is strikingly more affected by supranational rules than forest certification.

Just as labelling can be introduced in various settings, the debate on labelling can be enriched or suppressed in various ways. A consensus political culture can enrich a debate by allowing nuance argumentation, whereas an adversarial political culture allows the introduction of multiple frames. Other context elements reveal similar ambiguities. Sometimes an organizational landscape with big and few industrial actors facilitates labelling. At other times labelling is effectively counteracted in similar types of setting (Chapter 9). Skilfully and creatively relating to such context elements in the organizing and framing processes is critically important for improving conditions for labelling. Now, let us turn to these process factors.

# 8
## Three Framing Strategies: From a Complex Reality to a Categorical Label

### An introductory example: the moving and slippery nature of fish labelling

Labelling involves the challenging task of translating a complex reality into a categorical label. We understand complexity as including both a social and an ecological dimension. In part, we illustrated the social complexity in Chapter 6 by the many different arguments in favour of, and against, labelling. Within the broad battery of encouraging and sceptical arguments, there are divergent values, interests, motives, beliefs, and concerns. In a similar vein, ecological complexity could be illustrated by the many concerns that were raised in one particular labelling case: the Swedish seafood labelling. To what should fish and fishery with a green label refer? *First*, there is the concern about over-fishing, that fisheries and their current regulatory apparatus cause depletion or extinction of species and stocks. It was generally assumed that fish with a green label should come from healthy and sustainable stocks, that is, stocks that are within 'safe biological limits'. *Second*, certified fishery should minimize by-catches of other marine species, and the fishing methods should not cause harm to birds and seals or to the seabed. Then there is a *third concern* about how the boat engine emissions, fishing vessels, waste, and use of chemicals might damage the marine environment. A *fourth group of concerns* refers to landing and processing, including use of additives, waste, and energy consumption. Should a fish with a green label not be caught by a local fisherman, thus reducing transportation and discharge of greenhouse gases? *Fifth*, much fish contains high levels of toxins and heavy metals. Would consumers not assume that fish with a green label also refers to strict standards for such contaminations, even though the fishermen or the fish processing industry are not responsible for these contaminations? The fishery, processing, packaging, and distribution of fish products could still have many other effects on the environment. In addition, to push the eco-logical holism even further, should not fishermen on certified vessels eat organic food?[65]

Dealing with complexities requires dealing with diverse knowledge claims, ideas, values, and interests. Apart from the fact that various stakeholders have different encouraging and sceptical arguments about the labelling, they also have different experiences and expertise on themes of relevance for the labelling process. Practical experience as well as theoretical knowledge must be used. Again, the seafood label-ling case is illustrative. Obviously, fish is a moving resource, which implies that it is not easy to gain accurate knowledge and estimates of fishing stocks. Eco-labelling of seafood would be unthinkable without scientific advice such as the recommendations of the well-recognized International Council for the Exploration of the Sea (ICES) (on the role of this cognitive authority, see Chapter 10). Yet, fishermen often disagree with marine biologists and other scientists. These knowledge disagree-ments have much to do with the mobile nature of the fish resource. While scientists argue that fish stocks are declining, fishermen hold that the fish move depending on, for example, changes in water temperature. Many other experts advance *knowledge claims* as well, in relation to the labelling activity. Fishermen know how to fish and how to run a vessel, and they know whether certain requirements are economically and technologically feasible. Fish processing industries know what particular additives certain seafood products require. EMOs know what depletes biodiversity and damages the marine environment, and they know what is required by a credible labelling system. Authorities know how to inspect and how existing rules might impinge on the labelling. Retailers know what is tradable and what consumers ask for. Consumers know what they want, what they are afraid of, whom they trust, and sometimes what the maximum size of the holes in the fishing net should be.

The challenge does not refer only to knowledge claims being dispersed and difficult to unite. The challenge also concerns actors' mistrusting each other's knowledge claims and intentions. Scientists may believe that fishermen suggest alternative hypotheses only so that they can continue with their current fishing practice. In Chapter 10, we analyse the difficult task for a labelling project to deal with such mutual mistrust by trying to practise certain standard-setting ideals. In this chapter, we focus on framing.

## The concept of framing

Dealing with and uniting all encouraging and sceptical arguments, as well as motives and knowledge claims, requires that the various actors and stakeholders become active in *framing* the problems and solutions. Framing is an essential part of the translation of social and ecological complexity into a categorical label.

The definition of *framing* that Martin Rein and Donald Schön provide is instructive:

> [F]raming is a way of selecting, organizing, interpreting, and making sense of a complex reality to provide guideposts for knowing, analyzing, persuading, and acting. A frame is a perspective from which an amorphous, ill-defined, problematic situation can be made sense of and acted on. (Rein & Schön, 1993, p. 146)

Frames can be widely shared among a great number of organizations or they can be more specific to a certain organization. Actors refer to frames that are common in the general environmental discourse, for example, biodiversity, sustainability, and the precautionary principle – frames that are collectively recognized and used as a reference in communication about environmental issues. Thus, framing occurs in a discursive context (cf. Steinberg, 1998; Triandafyllidou & Fotiou, 1998; Chong & Druckman, 2007), but organizations interpret them slightly differently and make specific combinations of frames so that they accord with their identities, activities, and priorities (Boström, 2004a). Green labelling organizations and networks may themselves have to create frames that guide the labelling process.

To assess criteria for standards, labels, and certificates, the actors involved have to translate such complexities and uncertainties into categorical statements about, for instance, consequences for the environment, humans, health, economy and social conditions among workers producing the products. Such translations require the inclusion of certain factors and the exclusion of others. Framing can thus be seen as this process – calculated or accidental, explicit or implicit – of translating and making sense of a multifaceted world. In Chong and Druckman's terms, 'framing refers to the process by which people develop a conceptualization of an issue or reorient their thinking about an issue' (2007, p. 104). Applied to our cases of labelling, every actor involved is also involved in such processes of framing. The task for us as researchers is to identify these processes, and to analyse them in a way that makes them understandable. Moreover, our task of frame analysis has the ambition of highlighting opportunities for clarification, improvement, and so forth, for the actors involved in labelling and for the research community. Even though terms such as 'improvement' are very much in the eyes of the beholder, we still argue that it is possible for us to suggest certain modifications based on the stated and underlying aims of the policy process.

Frame analysis pays attention to actors' potentially active role in the construction of interpretative schemes (cf. Swidler, 1986). Therefore, framing can be an intentional strategic and conscious activity in order to mobilize commitments and convince various audiences. However, framing can also occur without much reflection on basic premises. The policy analysts can use framing theory to analyse both the explicit frames that policy actors construct, and the more implicit and hidden assumptions and understandings (Fischer, 2003). Applied to labelling and standardization, we argue that framing has a double role. At the same time as framing should simplify complexity, it is also essential for identifying, acknowledging, comprehending, and reflecting upon the complexity. Framing should stimulate reflection and communication about the social and ecological complexity of the labelling activity. In this chapter, we address this through an analysis of three framing strategies – boundary framing, frame resolution, and frame reflection – using our examples of labelling.

## Three framing strategies

Three framing strategies permeate the effort to establish and promote a green label.[66] Firstly, labelling actors use framing to determine where the border should be drawn between products and production processes that should be labelled 'green' and 'conventional', and to convince more actors to see the benefits of a certain labelling scheme. This is a construction and marketing process strategy for which we use the concept of *boundary framing*. Secondly, frames can help groups to make agreements, and, applied to labelling, to decide on the criteria and standards of a certain label. We will use the concept of *frame resolution* to refer to the possibility of resolving disagreements and controversies, thus establishing a plain label – again, against the background of diverging knowledge claims, arguments, and motives. Frame resolution is the process whereby the different actors involved in labelling develop a common understanding, a reciprocal idea, about the purposes and their concerns about green labelling. A degree of frame resolution is necessary because green labelling is a process in which uncertain, dispersed, and complex knowledge as well as diverging values and interests are translated into a simple, plain, categorical label. Thirdly, framing may stimulate reflections and dialogue on various aspects of relevance to the labelling. The concept of *frame reflection* denotes the possibility of improving understandings, and the degree of self-reflexivity of the

*Table 8.1*   Three framing strategies in labelling

| Concepts | Boundary framing | Frame resolution | Frame reflection |
|---|---|---|---|
| **Role of framing in labelling** | To draw – or move – the line between 'green' and 'ordinary' products and production processes | To resolve conflicts and turn diverging views into a uniform label | To increase clarity and openness of labelling schemes |
| **Examples** | Perfection vs. imperfection Precaution vs. 'yes-unless' Framings of biodiversity, naturalness, cleanliness | Through conflict: temporary frame resolution across opposing frames by agreeing on a third, external frame Through consensus: continuous frame resolution by using an eco-pragmatic metaframing | Intraframe reflection Interframe reflection |
| **Comments** | Framing should not be presented as a pure, scientific process, but rather as social, political, nevertheless crucial to environmental outcomes Framing has been highly successful in several cases (in quantitative terms: more products, broader market segments). Yet, it is a delicate issue where the lines should be drawn to avoid watered-down or too marginal labelling schemes | Can be achieved in many ways, sometimes rapidly. Still, more long-lasting and engaging resolutions need room for frame reflection and deliberations | Is typically given modest attention and effort in the other strategies. Still, to facilitate for frame reflection is instrumental in gaining reflective trust of consumers and the public |

debates. A challenge for stakeholders and labelling agents is to shape the main frames so that they are sufficiently concrete to stimulate rich reflections and the perception of many different aspects. At the same time, the framing should attract many stakeholders, and lead towards agreement regarding risk-reducing policies. All three aspects are, we argue, essential ingredients in a powerful green consumerism (see table 8.1). They partly overlap, and they may co-develop. However, there is also a risk that they conflict in some important respects with – or even supersede – each other, something that we will discuss in the concluding part.

## Boundary framing: to draw – or move – the line between the 'green' and the 'ordinary'

The concept of *boundary framing* (Hunt et al., 1994; Silver, 1997) denotes processes where movements and counter-movements construct their separate framings, often as 'good' versus 'bad' or at least as two distinct categories, that is, 'qualified for labelling' versus 'unqualified for labelling'. All labelling involves boundaries and distinctions: symbolic differentiation. Something has to be defined as green and something has to be defined as being located outside the green spectrum (sometimes explicitly as grey, black, risky, or unsustainable, or sometimes by not being mentioned at all). In the various labelling schemes there are at least as many boundaries as there are principles, criteria, and interpretations. However, non-labelled goods are not necessarily 'bad' or 'malignant'; they may be unfeasible to label for economic or practical reasons. The symbolic differentiation between labelled and non-labelled goods is nonetheless highly controversial, as was illustrated in Chapter 6 (e.g., labelling provides the market with misleading separations of identical products). Concepts such as *frame bridging*[67] and *frame extension*,[68] which are frequently found in framing theory, are also relevant to take into account in boundary framings. These concepts concern moving the boundaries outwards towards a more inclusive frame.

### To select the proportions of the 'green' and 'ordinary'

Labelling actors can use framing to determine the part of an entire industry that, in principle, should be able to get their products labelled. Should only the top companies – the 'real' forerunners – be able to label their products or should the labelling scheme possibly include a much larger number of producers?

*Example of inclusive vs. exclusive framing in green mutual funds*

In the case of green mutual funds, this question of green proportions is particularly relevant. A plethora of criteria and investment ideals can be found here: 'Light greenness', 'best in class' (e.g., the least polluting truck companies), as well as 'the potential of significant environmental improvement', are only a few of the principles through which boundaries are drawn between green and conventional investments. An informant on green investing maintained that this diversity is beneficial for the business as well as for investors' choice (despite other frequent calls for uniform standards).

Certain interest groups, for instance EMOs, tend to stress that only a top proportion of a business can legitimately be signalled as environmentally friendly (cf. Erskine & Collins, 1997). The labelling, in their view, should differentiate the best examples – perhaps 15 or 30 per cent – in an industry from the rest. Ideally, this brings dynamics into the system. Accordingly, the 'rest' will try to match the 'best', and, when a larger number of actors reach the standards, these standards can be sharpened even more.

In contrast, other actors, often within the industry, may favour inclusive labels and standards. According to them, it is important that more than a minority of market players have a realistic chance to label their products, and to gain the many business advantages that labelling may offer. When the American forest trade association, the AF&PA, launched their FSC-competing programme SFI, they framed certification as a common industry issue, rather than as an issue for individual companies (Cashore et al., 2004, p. 101). The certification should be for the industry, not for single companies. The following example also illustrates how actors use framings to counteract differentiation.

*Example of a framing to downplay the differentiation between labelled and non-labelled goods altogether*

The USDA, which controls the organic standardization, refuses to frame the organic label as particularly beneficial to the environment, animal welfare or health compared with conventional agricultural methods (whereas such a difference between 'conventional' and 'organic' – in terms of environment and animal welfare – is more taken for granted in the European and Swedish context). In the United States, the USDA rather frames the organic label as 'ideological' and as 'satisfying a specific niche of consumers' (Klintman & Boström, 2004). Also, in the controversy as to whether or not a mandatory GM label should be introduced in the United States, the opponents of a mandatory GM label frame it as a completely arbitrary and ideological construct (as merely a separation of 'substantially equivalent' products), whereas the proponents of such a label frame it as more or less pure science, as reflections of 'knowledge on which to base rational choices'. (Klintman, 2002b, p. 81)

## Perfection vs. imperfection

Actors may use framings of *optimization* – or even perfection – for protecting an exclusive label, that is, to ensure that green-labelled products reflect the optimal (not only the best available) alternative, an option that is truly sustainable, environmentally friendly, and so on. From this point of view, a label could be marketed as reflecting 'sustainability' rather than 'towards sustainability'. Stakeholders sometimes reject such a position, however, since it relies on an (arguably unproductive) epistemological position. The position contends that, because labelling criteria and technical methods for separating products are less than perfect, it is useless to try to separate products in the first place (i.e., a position of 'epistemic absolutism'). Still, the cases of both organic labelling and mandatory GM labelling reveal several denotations of such an epistemic absolutism, through calls for perfection:

> This [GM risk to health and environment] is a distinct class of risk which is directly associated with the process by which genetically engineered foods are produced. Thus, foods carrying this class of risk can be easily identified, based on the process by which they were developed – genetic engineering. In light of this, it is only fair that consumers be informed of this class of risks and thereby be allowed to exercise their own judgment as to whether or not to accept that risk. In short, genetically engineered foods should be labelled as such. (Fagan, 2000)[69]

Although key actors may be tempted to use framings of optimization when marketing their products, they often disregard optimization in the dialogue among themselves. All actors that participate in a standard-setting process with the intention of giving constructive input typically reject epistemic absolutism. The reason is that such a position is not a constructive platform for making compromises and reaching agreements. It proves to be impossible to claim perfection – or even optimization – given ecological and social complexities and because of great uncertainties. The fact that stakeholders eventually come to an understanding that perfection or optimization is not achievable does not mean that a greater audience, including consumers, has rejected perfection. Excessive trust may be based on a belief in perfection. If labelling agents instead believe it is a good idea to stimulate a more *reflective* trust, they should avoid framings of perfection.

## The precautionary principle vs.
## the 'yes, unless' principle

*The precautionary principle* may be helpful for actors eager to keep the boundary closer to a certain core. From the *'no, unless' position*, which is arguably more common in Northern Europe than in the United States, the label should be exclusive until science has 'proven' that certain processes are 'risk-free'. For example, restrictions of additives and chemicals in organic food are often motivated with reference to precautionary thinking. While it is difficult to find strong scientific evidence for the statement that organic food really constitutes healthier or safer options, organic labellers maintain that science may be unable to conduct such empirical testing due to the multifaceted nature of the issue. Precautionary framings can be confronted with the opposite reasoning. In the American cases, the use of the *'yes, unless' position* contends that eco-labels ought to be inclusive, until science has 'proven' that certain processes are risky. In the United States, this is the very basis of the FDA's refusal to require a mandatory label of GM food or ingredients:

> The [Food and Drug] Agency is not aware of any information showing that foods derived by these new methods differ from other foods in any meaningful or uniform way, or that, as a class, foods developed by the new techniques present any different or greater safety concern than foods developed by traditional plant breeding. For this reason, the agency does not believe that the method of development of a new plant variety (including the use of new techniques including recombinant DNA techniques) is normally material information within the meaning of 21 USC paragraph 321 (n) and would not usually be required to be disclosed in labelling for the food. (US Food and Drug Administration, 1992)[70]

Moreover, this 'yes, unless' position was the basis for the USDA's initial proposal that 'organically grown' GM food should be allowed (which was later withdrawn; see below on this debate).

## Framings of other general
## principles vs. their opposites

A third way to construct boundaries is by using general framings, for instance *biodiversity, cleanliness*, and *naturalness* (Klintman & Boström, 2008). General framings allow for some interpretative flexibility as to what criteria and processes should be included. At the same time, such framings offer cognitive platforms that cannot include everything. For example, the biodiversity framing quite effectively disqualifies large areas of clear-felling, not only for aesthetic reasons (a traditional concern in forest practices) but because clear-felling reduces the variety of species in the area. The naturalness framing in organic food labelling effectively disqualifies the use of artificial fertilizers, pesticides, additives, GM, and so on. The cleanliness framing in green electricity effectively disqualifies coal-based electricity. However, the interpretative flexibility of each of these framings implies that contentious issues, from an environmentalist point of view, may easily enter the debates. Counter-initiatives also use the language of biodiversity and sustainability. In green electricity, labellers have been concerned that the cleanliness framing does not effectively exclude nuclear power, as the following example shows.

---

*Example of a framing around general principles: cleanliness of electricity generation*

The framing in electricity with a green label has been intriguing. In Sweden, as in many other countries, a main frame of electricity has been 'cleanliness', in close connection with the discourse of climate change. Yet, there have been many debates about the labellers' interpretation of this frame. For example, in recent years there have been an increasing number of claims in the media and by politicians in Sweden, as in many other countries, that nuclear power is 'clean'. Moreover, the public does not normally view nuclear power generation as unclean in the way they view fossil-based electricity. In the policy discourse where climate change is treated as the main issue it has lately been more difficult for the eco-labellers to get cultural resonance for a frame that excludes nuclear power but includes, for instance, certain biofuels. To many people, biofuels appear to be more visible as a problem (with their impact on forest biodiversity and direct emissions, despite the stated zero-sum emissions). In addition, a debated phase-out of nuclear power to many people means an increased import of coal-based electricity from abroad, something which a large part of the public in several countries perceive as entailing increased ecological risks. A manifestation of this critical view within the frame of cleanliness in Sweden was that a couple of electricity companies a few years ago offered customers the chance to order electricity generated by nuclear power only – at a price premium. The 'greens' who preferred nuclear power to other electricity sources got this offer. This shows that the work of maintaining an attractive frame that also achieves the end of reducing a broad range of risks has not been entirely successful. On the other hand, the electricity with a green label scheme in Sweden has clearly helped to stimulate intensive debate on the issue. (Klintman & Boström, 2008)

## Frame resolution: to resolve conflicts and turn diverging views into a uniform label

Labelling requires agreement among a broad group of stakeholders, although opinions differ as to how broad this group of stakeholders ought to be, and how permanent the agreement ought to be. To reach agreement, a form of *frame resolution* is necessary regardless of the policy context of the labelling.

In several of our cases, existing frames, such as *sustainability* and *biodiversity*, have helped groups reach agreements. Actors use the frame of biodiversity, which, as we saw above, is employed to draw boundaries, but this frame has also been instrumental in frame resolution. It receives great resonance among scientific, political, and environmental movement circles. Biodiversity loss (because of excessive use and exploitation of natural resources, acidification, trade with endangered species, etc.) is increasingly regarded and legitimized as a global *environmental* problem, and received its own convention during the UNCED in Rio 1992 (Porter et al., 2000; Hannigan, 2006). To the extent that the frame of biodiversity gains such a legitimate status globally, it is also extremely important as a reference point from which stakeholders in forestry, fisheries, and agriculture can reach agreements. However, because of their interpretative flexibility, frames such as the one about biodiversity are not, of course, self-sufficient for settling disputes (other factors such as organizing processes and power play important roles).

There are different pathways towards frame resolution, both through conflict and through consensus dialogue. In a context marked by profound controversy, dialogue towards consensus by way of the prevailing frames may be an unlikely project to carry through. Yet, temporary frame resolution is still possible, by linking to an external frame that opposing parties may agree on. We illustrate such temporary frame resolution in the following example, in which a frame around consumer democracy proved to be important.

---

*Example of resolutions across separate frames: the debate about 'organic GMs' in the United States*

The US debate that has become well-known under the name 'The Big Three' is particularly useful for illustrating our points. Although this debate is well-described elsewhere,[71] it deserves to be mentioned that the USDA in December 1997 proposed that processes such as irradiation, sewage sludge, and – most controversially – genetic modification should be permitted under the organic label. The reasoning within the USDA contended that excluding these (unsynthetic) production processes (not 'proven' to be unsafe) from the organic label would falsely imply that the Department assumes that these three types of processes are less safe than processes permitted under the organic label (see also Klintman, 2002b). The united and intensive mobilization including the overwhelming public response via the Internet – the USDA received 275,603 comments – led to a withdrawal of the Big Three. The head of the Department at the time, Dan Glickman, who is strongly in favour of GM in food production, admitted that 'The response was 20 times greater than anything ever before proposed by the USDA' (*St. Louis Post Dispatch*, 26 March). The Big Three debate was initially based on a conflict between two separate frames, one founded on a frame of naturalness combined with a pragmatic and precautionary approach, and another pro-business frame of technological optimism. Subsequently, the controversy was reframed into a democratic issue – about free consumer choice – which no party dared to deny completely after the strong public reactions over the Internet. The argument that labelling empowers consumers – for inherently democratic reasons if not for environmental ones – at least led to a temporary settlement of the controversy.

In contrast to temporary frame resolution, *metaframing* is a way to achieve permanent frame resolution. We understand metaframing as the development of frames *across opposite poles*, such as natural vs. artificial or orthodoxy/stringency vs. pragmatism (Klintman & Boström, 2004).[72] If successful, metaframing can lead to the controlling of the main part of the discursive space surrounding an issue or field, and the gaining of frame resonance among large parts of the public.[73]

What is common in green labelling is the development of what we call an *eco-pragmatic* metaframing. In this, participating stakeholders recognize the value of both a *market-pragmatic pole* and a *pole of stringent EMO values* and they agree that both poles should have a significant impact. Some degree of eco-pragmatic attitude may be necessary in all kinds of green labelling, because the strategy is *market-based* at the same time as it relies on the granting of legitimacy and credibility based on *green principles*. Yet, many business actors do not automatically recognize the value of stringent environmental goals and criteria; and EMOs may not always be inclined to recognize market conditions even if they affirm market-based strategies. For example, stakeholders in organic labelling in Sweden – in contrast to organic labelling in the United States – are much more willing to recognize each other's intentions and standpoints, and they often agree on basic purposes and concerns in the labelling. In the United States, the policy debates and mobilizations are typically of a more adversarial character, which often leads to temporary agreements on labelling criteria. The participants in these debates are not inclined to recognize the legitimacy of the other pole (see also Klintman, 2002a, b). Therefore, the crystallization of an eco-pragmatic metaframing is a matter of degree that varies from case to case.

In organic labelling we find opposites such as 'the natural' and 'the artificial' competing with each other. Organic almost by definition prioritizes the natural before the artificial. It would be wrong, however, to say that the artificial is completely absent. Too much emphasis on strictly *natural* products may constitute a strict barrier for the scope of the labelling (in addition to the fundamental difficulties involved in trying to define what is natural). The market-pragmatic pole has pressed for allowing modifications of products so consumers could perceive them as 'normal'; labelled products must be accessible, marketable, and familiar (with the right colours, smell, shape, etc.). For example, sugar should be white (and not naturally brown), milk should be homogenized, and meat should have a nice rose-coloured shade (which requires the additive nitrite).[74] Devoted organic players, normally reluctant to processing and additives, have learned to live with such market-pragmatic adaptations. The relatively recent frame, 'organic processing', is a concrete expression of the struggle to find a balance between stringent EMO values and market pragmatism. For the purist, organic processing may be seen as an oxymoron, a contradiction in terms – since organic processing refers to both originality and artificialness. The term has led to debates about what ought or ought not to be included in the organic labelling scheme. In what way, and how far, can an organic primary product be transformed and still be organic?

Through metaframing, frame resolutions become institutionalized and practised repeatedly. When stakeholders suggest new product types that should be subject for labelling, the existing metaframe will immediately shape the discussions. A metaframe certainly does have boundaries. It cannot cover all existing viewpoints because in that case one of the opposite poles would cease to have any impact. Each pole sets certain limits for frame extensions on the opposite side. Boundaries become institutionalized and materialized in rule systems and organizational procedures. Matters that used to be subject to boundary-making disputes become increasingly taken for granted, and this can make frame reflection difficult (see next section).

In the introduction to this chapter, we referred to the seafood labelling case to illustrate all the complexities confronting labellers. How was frame resolution reached in this case? It was far from possible to accommodate all concerns and suggestions about this complex issue. As the seafood labelling case was run by an established labelling organization (KRAV), which had institutionalized an eco-pragmatic metaframing in its normal labelling activities, it is no accident that such a framing helped to resolve the complexities. Great effort was expended to meet a pragmatic framing of feasibility by the industry–business side, although some of the stringent EMO values were not compromised (Boström, 2004b, 2006a). The actors in the labelling project chose to prioritize marine ecosystem issues (e.g., safe biological limits, damage to the marine environment) while playing down environmental aspects regarding the vessels, the distribution, the processing, and toxins and heavy metals in seafood, something that concerned many consumers. The impact of the market-pragmatic pole was also evident in a kind of free-trade and technological optimism, which implied that all kinds of fishing methods might in principle be permissible, unless they damage the marine environment (see Chapter 7). A kind of trust – perhaps excessive – in the auditing function (see Chapter 10) is also embedded in this market-pragmatic pole.

## Frame reflection: to increase the clarity and openness of labelling debates

The third and final strategy, frame reflection, may be the basis for shifts that enable resolutions of intractable policy controversies (Schön & Rein, 1994) or the mobilization of more supporters to the policy project. Reframing may occur with or without frame reflection, thoughtfully or thoughtlessly (ibid.; Fischer, 2003, pp. 144–147). A crucial issue is to what extent labelling activities and the framings surrounding them can co-develop with reflections on themes that are directly or indirectly of the highest relevance to the labelling activity, and whether or not these reflections are widespread among a much broader public than just the key stakeholders involved in the labelling. Boundary framing and frame resolution both gain from frame reflection, whereas it is possible or likely at the same time that these framing strategies will get in the way of frame reflection.

If labelling is to stimulate such reflections, it has to go far beyond the simple goal of creating broad simple trust (excessive trust). The rhetoric about 'neutrality', 'objectivity', 'independence', 'dispassion', and 'expertise' that proliferates in labelling is unlikely to improve frame reflection. Audit reports and labels based on such framing devices 'give off' information rather than stimulating communication and reflection (Van Maanen & Pentland, 1994, p. 54; cf. Power, 1997). Yet, labelling and the framing surrounding labelling may also be a way of stimulating critical reflections, among both key stakeholders and the public. We have two variants especially in mind. In a previous publication on labelling activities (Klintman & Boström, 2008) we distinguished between *intraframe transparency* and *interframe transparency* to analyse two ways of making invisible risks much more visible than a simple label on products would manage by itself. We extend our discussion here by distinguishing between *intraframe reflection* and *interframe reflection*.

## Reflection *within* a frame

Reflection *within* a frame (or *intraframe reflection*) concerns reflection on substances and practices that do, or do not, fit within the established frames to which one subscribes. Hence, a frame can provide actors with a cognitive tool to reflect on various topics. A specific frame (precaution, biodiversity, naturalness, cleanliness) may enable stakeholders and people to perceive and understand things in novel ways. Frames contribute to new and/or changed attitudes, knowledge, and pictures of the world (Boström, 2004a).

For example, while the traditional framing of touched/untouched nature was the basis for much protest about the exploitation of the visual landscape, the biodiversity framing stimulates thoughts and discussions about a much broader range of corporate practices. On the one hand, the biodiversity framing may connect with traditional nature conservation values (richness, purity, wilderness, originality) and thereby gain a strong cultural resonance. On the other hand, the biodiversity frame may stimulate reflections on nature protection with a much wider scope than would be possible within the traditional framing of touched/untouched nature. Informed by the biodiversity framing, nature protection no longer merely concerns distant untouched areas, but the entire territory, the links between landscapes, and the ordinary management of natural resources (Rémy & Mougenot, 2002; Parviainen & Frank, 2003). Still, it allows for the continuous focus on endangered species and visual landscapes, thereby being a potentially fruitful frame also for protests based on aesthetic concerns (cf. Boström, 2007). In short, the biodiversity framing may stimulate broad reflection and communication – on both cognitive and aesthetic premises – on how to design forest, agriculture, and marine practices in the best ways.

*Example of intraframe reflection through the biodiversity framing*

Particularly in the forest labelling process, the dialogue and the negotiations among stakeholders proceeded from the biodiversity framing. The frame helped to visualize certain problems in forestry. Informants from forest companies agreed that traditional, large-scale forestry disturbs many eco-systems. It causes extermination or threat of extermination of many species, and leads to an excessively homogeneous landscape. However, while no participant in the forest labelling process questioned or tried to reframe what the labelling referred to, they debated quite intensively whether or not certain methods such as chemical fertilizers and the planting of certain foreign types of trees really cause depletion of biodiversity. It is worth noting that such methods were not banned from the standard, leading Greenpeace, for instance, to the decision to withdraw from the process. Yet, Greenpeace did not question the underlying biodiversity framing that guided the labelling as such.

## Reflection *across* separate frames

Intraframe reflection may lead to incremental changes in the general frame, but it does not address the questioning of the given frame as such. Reflection *across* separate frames (or *interframe reflection*) goes further by enabling critical reflections on the value basis and the usefulness of the frame itself. 'Is the dominant frame too narrow or too broad? Do our own frames in fact contribute to the problematic situation? Should we perhaps completely change the dominant frame that has been used to shape the tools for green consumerism?'

Whereas intraframe reflection may provide avenues for some extension of a frame, interframe reflection enables self-critical frame reflection and can therefore induce more powerful shifts. Such shifts require the existence of external frames from which the prevalent frame is interpreted, compared, and assessed. Biodiversity is seen in contrast to the original framings of untouched/touched nature, and this reframing stimulates reflections on a much broader array of topics. Eco-modernity has been developed from radical ecology and 'limits to growth' discourses (Hajer, 1995), reflective and self-reflexive consumerism from alternative lifestyle, and 'the Natural' is contrasted with 'the Artificial'. In fact, all frames require counter-frames as reference points.

Interframe reflection is consequently the capability to scrutinize both one's own frame that underlies an activity and another actor's frame, for instance that of an opponent. The latter type of frame reflection may serve merely to reinforce antagonism or may be used for manipulative image and impression management (Schön & Rein, 1994, p. 39). However, interframe reflection may also train actors to see and understand the blindness and limits of their own frames, and how the existing frames contribute to the problematic situation in which they find themselves (ibid., p. 187). Dialogue assisted by interframe reflection should therefore be fruitful for helping groups to reach agreement and end stalemate. Yet, such productive frame reflection requires both readiness and capacities among the participants:

> The process of frame-reflection depends in particular on the orientations of the participants: their relative distance from their objects under consideration, their willingness to look at things from other perspectives, their propensity toward 'cognitive risk taking' coupled with their openness to the uncertainty associated with frame conflict. (Fischer, 2003, p. 146)

Do we see any instances of comprehensive reflection across frames in the public debate in our cases? Yes, we do indeed see evidence of such interframe reflection in all labelling activities, albeit with varying levels of intensity, clarity, seriousness, and systematism; and interframe reflection may coexist with less advanced forms of thinking and dialogue.

---

*Example 1 of reflections across frames in the public debate: eco-labelling of electricity in Sweden*

In Swedish debates about the reduction of ecological risks by green labels, 'green' electricity is one of the most contested and adversarial. There used to be several energy companies that tried to present their windmills or hydro plants in a conspicuous way, thus making their 'clean' electricity generation visible to the public. However, almost any case of green electricity generation in Sweden is in fact subject to interframe visibility and reflection. In heated debates, each mode of electricity generation is confronted with competing frames, for instance environmentally oriented frames of 'visual and auditive aesthetics', 'local wildlife preservation' (where windmills are claimed be noisy and to kill birds), as a threat to 'biodiversity', or to 'locally valued ecological sites' (an issue which concerns hydropower and biomass in particular). Green electricity debates involve a broad range of diverging – often local – interests and ideals, not least in the name of greenness and cleanliness. It is fair to say that the green electricity label, through the frame-reflective policy discourse climate, has made the case visible and debated in a more fundamental sense than the label manages itself. In this way, a much broader range of risks and options are made transparent than those that are merely included in the manifest frame of cleanliness used by the actors who standardize the eco-labelling scheme. (Klintman & Boström, 2008)

---

Frame reflection in an intractable controversy may ideally have the advantage of bringing fundamental value differences and bottom lines up for open public debate, a level rarely reached in milder disagreements. Interframe reflection visualizes the boundaries of the given frame, and even how a dominant frame conceals alternative ideas.

*Example 2 of reflections across frames in the public debate: social key biotopes in forestry*

While the biodiversity framing can enrich reflections among both laypersons and participating stakeholders in a labelling project, it is also vulnerable to absorption within an expert-dominant approach. Indeed, few can go out and take a walk in a forest and make a quick statement about the state of the biodiversity – 'ordinary people do not "see" the biological diversity' (Peuhkuri & Jokinen, 1999, p. 135). We have to rely on experts, who in turn have to rely on theories, methods, models, indicators, maps, and so on. Strong reliance on the biodiversity framing may stand in contrast to other visions of the landscape based on aesthetic reflexivity (cf. Lash, 1994). A local protest group that was criticizing an FSC-certified forest company for its intention to fell a forest area tried to bridge the frame of biodiversity to a frame of *social sustainability* (Klintman & Boström, 2008). They framed their forest as a 'social key biotope' and as 'the last intact mountain' in the game protection area. However, the company, the authorities, and the certification body downplayed their efforts and concern for 'their' beautiful forest, for outdoor life and tourism in the area. This interframe debate reveals an expert-dominant and eco-modernist discourse, which ignores the evaluation based on laypersons' living experiences. The operations of the certified company were seen as already legitimized by the FSC standard, other state measures, and indirectly by the biodiversity frame.

## Limited frame reflection in a policy culture of metaframing?

An *eco-pragmatic* metaframing, in which the participants recognize the value and needed impact of both the market-pragmatic pole and the pole of stringent EMO values, may not only have positive consequences (permanent frame resolution). As the frame, mostly implicitly, guides the dialogue and negotiations, it also sets limits on what it is acceptable to communicate and negotiate about. Certain matters are typically excluded systematically from the discussions. A rigid eco-pragmatic framing may be effective in permanently and systematically excluding factors, and not uncommonly factors that are seen as central and important in the general environmental debate. This has been apparent in the organic labelling case in Sweden, in which several topics – such as local food, transport or energy use in food production – were largely excluded from labelling discussions (Klintman & Boström, 2004; Boström & Klintman, 2006a):

*Example of limited frame reflection: principles of organic food labelling in Sweden*

Small-scale farming has been treated as an irrelevant aspect within the organic label. The socioeconomic – and some would argue ecological – risks of not giving small-scale farming a particular status are often neglected and made invisible through the eco-labelling of food. The same is true for global transports of food with a green label. As to transport, and its risks for climate change, this is defined as irrelevant to the eco-label. Moreover, the original association between 'naturalness' and the local realm is thereby obscured because it is incoherent with the eco-pragmatic framing. In fact, the problem is not only that labelling discussions exclude such themes (themes that are clearly relevant from the perspectives of environmentalists). Swedish debates on green consumerism in the food area have rarely included such topics. This is surprising because eco-labelled, or 'natural', food is an area with a strong historical connection to small-scale and local food production. It started as a critique of the increasingly 'industrialized' food production where food is transported around the globe despite the fact that local resources are available. (Guthman, 2004)[75]

## Conclusions: trade-off between the framing strategies?

Framing is essential in labelling. Framing denotes processes in which actors deal with social and ecological complexity. Stakeholders develop their arguments through frames, and these frames also help them find common grounds for negotiations and compromises. Both the standard-setting process prior to the introduction of green labelling and the subsequent use of the label may stimulate rich debates and reflection both within and across frames, making stakeholders and the general public able to develop arguments, debate and reflect critically on categorical statements. Used reflectively, frames such as biodiversity, cleanliness, precaution, naturalness, sustainability, and many others are themselves useful for perceiving and understanding the set of problems in novel ways. The multi-layered character of these frames may open up interpretative flexibility. The label itself serves as a tool and symbol, around which such frames are concretized, materialized, and practised. Optimization or perfection in either the label or the frame (measured, e.g., in terms of internal consistency) is seldom very beneficial for the frame reflection or for improving such values as *consumer insight* (cf. Chapter 4). Perhaps the contrary is closer to the truth. Consumers' observations of imperfections in the schemes and the frames could be a constructive basis for reinterpretation, something which could be of immense value for labellers and other policy actors.

Our informants, who represent certain stakeholders in the labelling processes as well as many other stakeholders (including laypersons), indeed express rich reflections both on the usefulness and limitations of labelling criteria and levels as such and on the underlying frames. At the same time, in none of our cases have the debates been entirely successful in stimulating such critical thinking and reframing for broad audiences. For example, the eco-pragmatic metaframing, which is particularly characteristic of the Swedish organic labelling organization, forcefully and systematically includes and excludes aspects. It has proven difficult to develop alternative visions 'beyond organic'. Other frames exist, but the dominant one tends to marginalize or absorb them. The problem is not that social and ecological complexity has been simplified, because that is an inherent part of both framing and labelling. The problem is rather that a particular simplification becomes fixed, not an issue for further discussion, reflection, and reframing. Nor are excessively polarized statements – as could be found in the American cases – beneficial for mutual frame reflection. Such frame debates are more likely to lead to confusion than to deliberative dialogue and reflective understanding.

The labelling process necessarily involves systematic inclusion and exclusion of aspects, but the process is problematic if the debate climate does not give room for open and balanced reflection and criticism of a frame. The framing processes therefore have drawbacks in stimulating the kind of consumer engagement, insight and reflective trust that we believe are essential for the long-term ability of green labelling to contribute positively to a consumer society that is both greener and more democratic. Reflective trust relates to a multiple framing context in which consumers may utilize different frames in their assessment and use of green labels. However, it is clear that the strategies for boundary framing and frame resolution are generally treated as overriding goals compared with frame reflection. To be sure, green labelling stimulates 'green, reflective consumerism', but reflections (especially at the interframe level) within and about the labelling debates among a broad group of stakeholders occur unintentionally rather than on purpose. Based on framing theory, this last statement would hardly be surprising, but it is nevertheless interesting and highly relevant to discuss how to achieve a more systematically reflective and green consumerism. This is an issue that we have reason to return to later in the book. In the following two chapters we analyse the organizing process surrounding labelling.

# 9
# Organizing the Labelling

The previous chapter emphasized the significance of framing processes for the translation of a complex reality into a labelling scheme. This chapter turns the focus to another process in this translation, namely the organizing that takes place behind labelling schemes. For our purposes, it is relevant to employ and combine two perspectives on organizing: first, one broad perspective that focuses on coalitions, social movements, and interaction among actors within the organizational landscape; second, one narrow perspective that focuses on formal organization. A brief discussion of these two perspectives is followed by an analysis of different organizational forms for labelling activities. We analyse three different organized forms in which business and social movement actors, particularly EMOs, interact in labelling arrangements; and we ask whether and how such variation matters. Indeed, organizational form itself is subject to intense debate among stakeholders, because the stakeholders generally believe that organizational form matters both for the efficiency of interaction across actors and for the environmental as well as social outcomes, and not least for the legitimacy of the labelling scheme. Finally, this chapter takes a closer look at some of the major actors involved in labelling, including their motives (or lack of motives), arguments, and power resources.

## Two perspectives on organizing: coalition and formal organization

The broad perspective on organizing processes would, in part, focus on the mobilization of larger coalitions, *advocacy networks* (cf. Sabatier & Jenkins-Smith, 1999), and *social movements* (McAdam et al., 1996). Social movement mobilization, media protest, and coalition-building are crucial steps towards the establishment – or the counteraction – of labelling projects. In studies of environmental politics and regulation, scholars often explain reform inertia by reference to one dominant iron triangle with actors representing an industry, labour unions, and state agencies (e.g., Lundqvist, 1996). Green labelling is no exception (e.g., Hofer, 2000; Elliot & Schlaepfer, 2001). Political consumerist strategies such as green labelling are often preceded and paralleled by other, more confrontational, social movement strategies, for instance media protests and consumer boycotts.

From a power perspective, coalitions are primarily visible threats to prevailing power structures in the organizational landscape. Increasing criticism and negative publicity addressed towards a particular business can suddenly change the willingness among business players to consider voluntary regulation (Boström & Garsten, 2008).[76] Seen in this way, there is a reputation risk for companies if they ignore initiatives from well-recognized SMOs aimed at contributing to positive political consumerism. Coalition-building involves disseminations of joint messages to a wider audience about the need for urgent change. Companies are more likely to legitimize green labelling schemes if the current practices of the companies are considered a major problem in society (Bendell, 2000; Cashore et al., 2004; Holzer, 2007). In addition, opportunities for symbolic benefits are created by accepting being partners in new coalitions.

Building new coalitions that successfully challenge old coalitions and power structures is difficult, however; it is a resource-consuming task. It takes time – often a decade or so, according to Sabatier and Jenkins-Smith (1999). In our labelling cases we also see that new coalitions can easily be counteracted by other coalitions. Coalition-building in the case of eco-certified forestry, globally as well as nationally in Sweden and the United States, which we described in Chapter 5, illustrates well the critical role of both coalition-building and counter-coalitions.

In the United States, well-represented and unified companies within the AF&PA effectively counteracted the social-movement-led FSC initiative by implementing a competing model. The AF&PA responded aggressively to the FSC initiative and successively mobilized support from public bodies, states, and retailers:

> The AF&PA is important for our story because it provided an immediate organizational setting in which the industry could undertake proactive and strategic choices in response to the increasing pressure for forest certification. (Cashore et al., 2004, p. 95)

Accordingly, this well-unified industry could mobilize resources, expertise, external coalition partners, strategic skills, and arguments to defeat labelling initiatives from EMOs. On the other hand, a well-coordinated industry can be a useful coalition partner also for labelling initiators. In Sweden, a well-unified forest industry actually joined the FSC, and participated in its promotion and implementation (see also Cashore et al., 2004). Although some forest companies were clearly hesitant about the FSC initially, an EMO-led coalition was able rather quickly to engage the entire Swedish forest industry in the FSC standardization process. Several forest companies were critical at first, but, when EMOs were able to convince a couple of big Swedish forest companies to commit to the FSC working group, other forest companies were triggered to join. It mattered a lot that the Swedish forest companies had a long tradition of communicating internally within the industry on common matters through the Swedish Forest Industries Association. An organizational landscape with strong industrial and producer-based associations can therefore be instrumental in both effectively supporting (Sweden) and counteracting (United States) labelling initiatives.

However, coalition-building should not merely be seen from a power perspective, that is, as power struggles based on given interests. Another important dimension of the coalition-building process is the gradual development of common understandings, interpretative frames, and even shifting interests among actors that were initially antagonists. From this perspective, coalition-building is especially favourable for frame-bridging strategies. An interesting point in existing theories of coalition-building (Hajer, 1995; Sabatier & Jenkins-Smith, 1999) is that language, discourse, and dialogue may fundamentally alter the beliefs, interests and values of the actors involved.

A shortcoming of these theories, however, is that they neglect the role of formal organization. As we see it, framing can help bind actors together; but it should be emphasized that this is done within concrete policy networks, in which actors orient their strategies and actions towards each other. Indeed, one concrete labelling project with clearly defined goals may in itself constitute a link that helps actors relate to one another, a link that facilitates the development and bridging of frames, and even the development of mutual reflective trust. The above-mentioned theories fail to ask a crucial question: what does the establishment of new formal organizations mean for the policy process? In activities such as standardization and green labelling, this question appears to be important. Whereas coalition-building may be an organizational basis for an initial development of new frames – of common understandings, mutual learning, and mutual respect – the building of a coalition may only be a temporary step.

A narrower perspective on organizing processes focused on formal organization is a useful complement. It is relevant to analyse labelling through such organizational lenses because green labelling has become an organizationally institutionalized activity in environmental policymaking. It has become a permanent activity which requires resources, administration, and enduring, rule-based interaction between parties. Running activities permanently normally requires a formal organization. Establishing a formal organization can itself be helpful in carving out a permanent position in the 'regulatory space' (Hancher & Moran, 1989). Having institutional status as a formal organization is a basic symbolic resource (Brunsson & Sahlin-Andersson, 2000). Then the public and many other organizations in the social landscape can more easily recognize the labelling project. Establishing a formal organization is a signal to the surrounding society that 'we are here to stay', 'we intend to go further', and 'this is not a temporary project'. There are, of course, other equally important manifest administrative roles of a formal labelling organization.[77]

By tradition, organization theory has seen formal organizations as rather clear-cut entities, as is the case of societal spheres. The state, the market (which includes companies), and civil society (which includes categories such as voluntary organizations, NGOs, or SMOs) are all seen as relatively autonomous, having their own internal logics and dynamics. Several of the labelling organizations we analyse in this book are very difficult to categorize in such ways. Many of them operate as hybrid organizations in that they include actors that belong to more than one of these spheres. Hybrid arrangements reflect a standard-setting ideal of inclusiveness and broad representation (Boström, 2006b). Many labelling organizations are at the same time meta-organizations in that they have other organizations, not merely individuals, as members (cf. Ahrne & Brunsson, 2004b, 2008). Such patterns may create specific opportunities and tensions, as we will see.

However, the level of inclusiveness and a number of other features in labelling organizations is a matter of variation. Organizers of labelling consider a range of organizational issues, ranging from constitutional issues to everyday routines. There are hundreds of similar matters, many of them controversial. The most basic is the issue of whether the organization should be set up as an association, a foundation, or a private company, or whether a state agency should run it. Another issue is whether the standard-setter also should control the certification and accreditation. A third issue is what type of funding should be mobilized. A further crucial question, to which we devote a number of pages, is *whether* and *how* to include different categories of actors, for instance business actors and social movement actors.

## How to organize SMO–business interaction in labelling arrangements?

Which actors should be allowed to be members or participants in the labelling organization and share the power of decision-making? Which actors should act only in advisory capacities, and which actors and claim-makers should the labelling organization leave outside all such processes? Does such variation matter? If so, in what sense does it matter?

We are particularly interested in examining the interaction between business actors and social movement actors. It is difficult to conceive of labelling without, in some way, the inclusion of business actors, because labelling is necessarily related to products and production. Social movement criticizes and targets certain industries that are subject to green positive political consumerism. Much of the credibility of the labelling organization has to do with the involvement of SMOs. Among these, EMOs are important in the case of green labelling; but the involvement of other groups is also significant, such as animal welfare organizations, labour unions, consumer groups, religious associations, indigenous groups, and development groups. In the following analysis, we distinguish between three categories: SMO-governed labelling, hybrid-governed labelling, and business-governed labelling.

### Three types

**SMO-governed labelling**: in this case, the SMO fully controls the labelling. This appears to be rare, but from our cases it is exemplified by SSNC's Good Environmental Choice, which includes green electricity along with several other household products. SSNC is an individual member-based democratic organization; it is the individual members who ultimately decide on labelling priorities and criteria. Despite the fact that SSNC controls the labelling, it has to engage in dialogue with a range of other interest groups, for instance corporations that might be interested in labelling their products. Companies participate in advisory capacities. We return to this case below when comparing the different types.

**Business-governed labelling**: this is an opposite case to the previous one. Businesses choose to take care of the labelling themselves without consulting SMOs. They ignore ideals of stakeholder representation, dialogue, and inclusiveness of the criteria-setting for the labelling.[78] Instead they choose to develop labelling principles and criteria in a self-sufficient way. Self-regulation is preferred to co-regulation. That does not necessarily mean, however, that they ignore signals, claims, and messages from external stakeholders about what companies should do, and what should be seen as green products. They may relate to existing standards in order to strengthen legitimacy and credibility. SMOs and other actors can still play a role as claim-makers in the general debate. Yet, the business labellers simply want to control the agenda.

The AF&PA tried initially to develop SFI in this way but was forced to develop a more inclusive arrangement (see below). Another example is green trademarks, for example the Nordic food retailer Coop's Änglamark. In order to ensure credibility, Coop normally also uses the organic label (KRAV) for its Änglamark-labelled products. Yet, in specific cases when Coop disagrees with KRAV about standard criteria, Coop uses its Änglamark label on its products without the KRAV label. Thus, although Coop is a member of KRAV, there is simultaneously a symbolic competition between KRAV and Änglamark.

Business-governed labelling is the typical means of labelling in the area of green and ethical funds. In the SRI industry, certain well-established SRI investment companies launch their own social and environmental indexes. For instance, the American mutual fund company Domini has an index, in its own name: Domini Social Index. Although it is fair to assume that Domini has regular communication with SMOs, and is highly influenced by the general information that comes from non-profit organizations with green and social profiles, Domini tries to strengthen its name by developing its own index, albeit in close cooperation with another player.[79]

Green mutual funds – in Europe and the United States – are largely organized by the banks and the 'mutual fund industry' (Radin & Stevenson, 2006), that is, the private, for-profit sector, albeit consulted by the broader social investment movement including various screening firms and organizations (Schwartz, 2003). Whereas traditional EMOs still play a modest role in green and ethical mutual funds, several inventive constellations have been established that consist of business associations providing services for socially responsible investment institutions and financial professionals. The Social Investment Forum, with Co-op America as its secretariat, is an example.

Significant to business-governed labelling (perhaps mainly within the mutual funds industry) is the double-edged 'labelling effort'. On the one hand, they are eager to comply with principles of large, standard environmental and social indexes launched by organizations such as the UN and the Dow Jones Sustainability Indices (DJSI), partly to converge with a green and ethical investment movement. On the other hand, each mutual fund company is inclined to separate itself from competing mutual fund companies, by using its own, much less overt ethical and environmental strategies. It seems that competitors on the mutual funds market partly treat ethical criteria – the bases for exclusions and inclusions of companies – as business secrets (see also Chapter 10). It is also plausible that an involvement of environmental, social and consumer NGOs in requirements and scrutinies of the latter will be a key issue for the alternative mutual funds movement(s) for the near future.

**Hybrid-governed labelling**: such labelling is performed when a social movement or an assembly of SMOs shares the formal decision-making power with other groups and organizations. These hybrid arrangements are therefore normally meta-organizations (Ahrne & Brunsson, 2008). However, they may look very different. The decision-making power among SMOs can be strong, unclear, or weak. The FSC is the emblematic example in which SMOs have strong formal decision-making power. Indeed, one significant feature of the FSC is that the decision-making power is divided strictly equally among three groups of interests: social, environmental, and economic.

The FSC competitor SFI was created and run entirely by the trade association AF&PA with no external stakeholders involved (business-governed labelling). This model was deliberately chosen because of the strong criticism among the American forest industry of the FSC's organizational form, in which economic interests have only one-third of the decision-making power (Cashore et al., 2004, p. 109). Yet, they gradually reformed the organizational structure to respond to the criticism addressed by FSC promoters. The AF&PA created the Sustainable Forestry Board (SFB), in which new interests were given some limited formal decision-making power. Since 2002, one-third of SFB members are environmental community representatives, one-third are SFI members, and the remaining third consists of members of the broader forest community (ibid., pp. 117–118). Although most EMOs, including the major WWF and Sierra Club, still do not participate in the AF&PA, a few environmental and conservation groups eventually joined the SFB. A degree of inclusiveness is thereby created, which in turn seems to have given SFI more credibility in the American marketplace (ibid.).[80]

Still other hybrid organizations, such as the Nordic Swan and KRAV, have more flexible criteria regarding the allocation of decision-making power than the FSC, with the implication that the power relation between business members and SMOs is unclear and shifting.[81] By comparing FSC and KRAV we have seen that the former organization has more effectively prevented power shifts over time, whereas we see increasing dominance of retailers and processing industries in KRAV, which must be understood in relation to their different governance structures (Boström, 2006b).

## Does organizational form matter?

Variations in organizational form are subject to intense debate among the stakeholders when labelling is introduced and organizational forms are designed. The fundamental reason why interest groups debate intensely about the forms is that they believe that forms matter. The choice of organizational form for the labelling process may indeed affect the content of labelling principles, criteria, and indicators. If business interests cannot control the process on their own, they may fear that environmental interests set strict standards and thresholds, which would be costly to comply with. All other things being equal – and in agreement with intuition – more inclusion of environmental interests means more stringent EMO values, and more inclusion of business interest means more market pragmatism.

To give one comparative example, the relation between organizational form and standard content is assumed *a priori* in the two competing models of FSC and SFI. The FSC promoters deliberately designed organizational forms and procedures 'with a view toward eliminating business dominance and encouraging relatively stringent standards' (Cashore et al., 2004, p. 12). FSC uses non-discretionary, substantive rules, and presupposes a relatively high number of specific, mandatory, on-the-ground requirements. It uses procedural rules, such as management plans, to facilitate the implementation of substantive rules. The policy scope is broad, covering rules on labour and indigenous rights and wide-ranging environmental impacts.

Likewise, a fundamental reason why the AF&PA created the SFI was that companies in the United States were particularly concerned about the FSC's wide-ranging, performance based approach to forest management and its chain-of-custody requirements. They criticized the FSC standards for being increasingly stringent, not appropriate for industry, and uneven across state jurisdictions (as regional-specific FSC standards were developed across the United States) – resulting in, as they argued, 'an unfair playing field that was more the result of politics than ecological differences' (ibid., p. 112; see also Gale, 2004).[82] The SFI instead uses discretionary, flexible rules. It sets non-mandatory indicators that *can* be followed. It is only procedural rules that are mandatory. Such procedural rules are seen as inherent ends, reflecting a belief that they will result in decreased environmental impact. The policy scope is narrower, focusing on management rules and the notion of continual improvement. The similarities to another kind of eco-standard, environmental management systems, are obvious (see Chapter 2), although this forest standard includes some substantive rules.[83]

Accordingly, form appears to affect content, but can there be reasons to question or tone down the relation between form and content? Are labellers overly worried about the consequences of organizational form? Based on analysis of our cases, we see at least four reasons to tone down the strength of this relationship.

In the first place, context factors, such as political culture, existing regulations, and existing power relations in the organizational landscape always provide opportunities for, and limitations on, what type of form it is possible to establish. In a consensual political culture it is naturally easier to establish hybrid arrangement where groups engage in dialogue, negotiations, and agreements.

Second, competing eco-standards are to some extent mutually dependent. In forestry, the stringency of the FSC has some relation to the stringency of its competitors, and vice versa. They must differ – but not too much – in order to assure symbolic differentiation. Although the FSC and the competing arrangements (PEFC, CSA, SFI) continue to be competitors and even antagonists, scholars have observed that the systems have gradually moved closer to each other (e.g., Domask, 2003, pp. 177–178; Cashore et al., 2004; Cubbage & Newman, 2006, pp. 265–266).[84] In the organic case, we note a similar development. Although the organizational forms differ, the American and European standards have gradually moved closer to each other.[85]

The gap between the 'stringent' standard and the competing standard cannot become too wide. That is because companies that are certified according to the stringent standard may find the costs associated with following stringent standard criteria unfair in relation to competitors that can certify their products without these costs. Likewise, the competing standard cannot lag too much behind in terms of environmental stringency, since that would undermine its legitimacy.

Such types of competition and dependencies also appear between different types of eco-standards, such as between green labels and environmental declarations. This is evident in the case of paper. The Paper Profile used the same parameters as do the Nordic Swan and Good Environmental Choice (Nilsson, 2005). Hence, the companies using Paper Profile must document information using the same parameters as the competing labelling programme includes in the criteria document.

Third, EMOs, other SMOs or other expert organizations that are strong in terms of credibility and public recognition can also have great impact as advisors. Excluding certain actors from decision-making forums is certainly not the same as excluding their arguments. In the seafood labelling case EMOs were allowed only to have consultative roles, that is to say, no share in decision-making,[86] but the comments and viewpoints of such EMOs as WWF and SSNC were seen as critical, as was evident in our interviews with the other stakeholders involved, particularly those who managed the process.

The responsiveness towards SMOs among those who run the labelling programme is also important. According to Gulbrandsen (2008), the efforts of AF&PA to create a more inclusive organization that also involves other stakeholders outside the forest industries should be seen more as a strategic adaptation to popular ideas about legitimate organizing than as honest ambitions to invest capacity for responsiveness.

A fourth factor that weakens a strong relation between form and content is *the fact that the relation between business actors and SMOs is asymmetric*. Because the labelling strategy is a market-based policy approach that is dependent on market conditions, a labelling organization with one EMO as the sole principal, such as the SSNC's 'Good Environmental Choice', must in some way always include market-pragmatic reasoning and arguments, by making relevant compromises, and by using appealing, frame-bridging arguments. In practice, SSNC must stimulate a dialogue with a segment of relevant business actors, although SSNC has the final decision-making authority about labelling principles and criteria. A similar kind of market-pragmatist tendencies can be seen in the area of fair trade, which is also a strikingly SMO-led endeavour (Le Velly, 2007).

---

*Example of asymmetric power and market pragmatism of EMOs*

SSNC's decision to include under the eco-label all hydropower generated from rivers exploited before 1997 – the time when the label was established – has been perceived by actors within and outside the organization as a striking example of market-pragmatic reasoning and flat compromising. There were several motions contending that it is not correct to give the current hydro-power an eco-label since it goes against the struggle of the SSNC against the exploitation of Swedish rivers (*DN*, EKO, Lars-Ingmar Karlsson, 21 May 1996). Christer Nilsson, professor of ecology of rivers at Umeå University, for instance, claimed that SSNC's decision made the large energy company Vattenfall look clean again. As a consequence, he chose to end his membership of the SSNC. (*Svenska Dagbladet*, Inrikes, Susanna Baltscheffsky, 22 May 1996)

---

Although SSNC expresses a wish to sharpen its standards continuously, such a strategy can be risky because companies that have labelled their products might be unwilling to undertake further improvements, and they may have made costly investments which have to be paid off before new investments. For instance, between 2002 and 2003 there was a dramatic reduction (of more than 50 per cent) of companies with a licence proving that they used electricity with a green label, partially due to the stricter rules that the SSNC had introduced.[87]

Industry-dominated certification and labelling bodies do not necessarily have to make a corresponding adherence to the pole of stringent EMO values. At least in theory, they may label their products and merely pretend they are green; that is, greenwash. They can initiate the labelling without taking any notice of advice from SMOs. If they do this they are likely to face difficulties in convincing consumers and buyers about the value, credibility, and stringency of their label. Still, it can sometimes be enough for the competitors that important buyers view their labels as 'acceptable' – not 'best in class', 'but better than the average' (cf. Cashore et al., 2004). Looking at mutual funds, this can be exemplified by the distinction between 'dark green' and 'light green'. According to the British definition, the former partly refers to exclusions of whole industry sectors, such as oil and gas companies, from investment. The latter instead refers to 'best of sector' strategies, which may include investment in, for instance, oil and oil-based industry, but not in the worst polluters within such industries (yet not usually tobacco, armaments, etc.).[88] The broad popularity among businesses of another eco-standard – environmental management systems – shows how this argument may be perceived as rational.[89]

## Form, legitimacy, and inclusiveness

Organizational form is not simply designed for achieving a particular outcome, but must be seen in relation to all attempts at establishing credibility and legitimacy for one's activity, which is a topic that the school of sociological institutionalism emphasizes. For instance, whereas 'inclusiveness' can be seen as instrumental for the mobilization of dispersed resources among actors believed to be essential for the operations and for the labelling organization's market success, this standard-setting ideal can also be seen as of intrinsic value, connected to democratic values of deliberation and representativeness (Boström, 2006a, b).

Inclusiveness can also have a relation to other aspects that we emphasize in this book. An inclusive, hybrid organizational arrangement may have capacity for providing fruitful settings for the development of mutual reflective trust among a broad group of stakeholders. We will analyse the role of repeated interaction for establishing such trust in our next chapter. An inclusive arrangement may develop novel ways to achieve consumer representation, although consumer participation appears a challenge in all labelling arrangements (see below in this chapter).

Furthermore, an inclusive hybrid organizational setting is more likely to give rise to convergence among business and SMO actors, and it can provide an organizational platform for the development of an eco-pragmatic metaframing. An inclusive, hybrid-type labelling organization may – at least initially – facilitate the inclusion of alternative frames in the debates, which in turn leads to inter- and intraframe reflection in the early standardization work. A hybrid form of organizing labelling can facilitate communication and interaction among participating stakeholders. It provides a setting in which actors can contrast and interpret diverse sub-frames in relation to each other.

However, as interaction among the stakeholders gradually institutionalizes, the multi-frame character of the debate tends to diminish (cf. Dryzek, 1993, 2001), and interframe reflection is restrained. The form of inclusiveness may force actors to become too integrated in that the arrangement gradually suppresses multiple framings and frame reflection. Members of a labelling organization may feel strong expectations to express loyalty to basic labelling principles. Thus, a mutual expectation grows among member organizations that members should not express serious criticism in public. For example, as key actors in the agriculture and food industry become included as full members of

KRAV, they consequently must officially approve of organic production; they cannot officially question basic organic principles. This is not necessarily beneficial for the debate on organic production. Finally, extraordinarily well-balanced power relations may be inflexible and cause stalemate in discussions, negotiations, and decision-making.[90] If participants are only active as advisors, they lack some critical power means, but they do not have to take responsibility for actions, and they can more easily criticize the system later on, and support alternative frames and programmes.

By summarizing our analysis of the relation between organizational form and standards contents, we wish to emphasize that organizational form appears to be a highly intriguing and controversial topic in labelling and other standardization activities (see also Boström & Garsten, 2008; Tamm Hallström, 2008). Often the debate among stakeholders concerning organizational form relates to the outcome of standardization, that is, the stringency or flexibility of the standards. Yet, as we have seen, the labelling content is not simply a reflection of the chosen organizational form. The form can affect the balance between stringent EMO values and market pragmatism, although the form cannot alter the need to find such a balance. We would nevertheless hold that the choice of form is critical for labelling initiators to consider (for labelling output, for business vs. SMO control, for development of reflective trust, for consumer representation, for positive and negative aspects of inclusiveness, etc.). SSNC can control its standardization agenda, with the help of its internal democratic procedures for decision-making. Its own members can raise concerns about far-reaching compromises with industries. The members can mobilize and say stop. If SMOs share decision-making power with business groups, the SMOs can do more than just voice their concern. Within the FSC, non-profit groups (environmental and social interests) can always unite and use their veto. Hence, such formal decision-making power can serve as an antidote against dubious standardization projects, and against greenwashing of industries. In contrast, not including SMOs is a guarantee for not including certain criteria in the standards.

## Actors involved: interest, roles, and power

This section takes a closer look at key actor categories within labelling arrangements. The figure 9.1 below shows typical actors that potentially can be involved in labelling. There is not enough space to discuss the role of all possible actors here. Instead, we will concentrate on retailers,

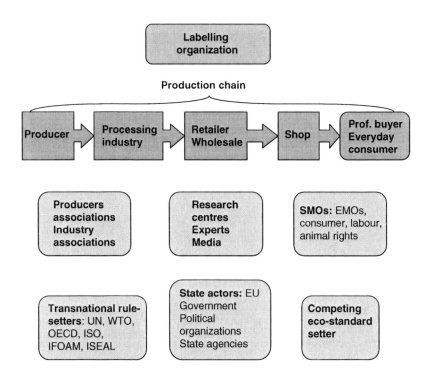

*Figure 9.1*   Actors involved in labelling

producers, social movement organizations with a focus on EMOs, and finally the issue of consumer representation and participation. The roles of transnational and state rule-setting actors were examined in Chapter 7, and we will discuss the role of cognitive authorities in Chapter 10.

### Retailers and processing industries

A recurrent pattern in several of our cases, except our paper and green fund cases, is the central role of business actors in the middle of the production chain, particularly retailers but also processing industries. This accords with recent literature which stresses that the power of retailers has increased significantly in recent decades. A retail-led form of food governance has emerged, according to Marsden and his colleagues (2000; Klonsky, 2000; see also Nestlé, 2002). This is due to factors such as

- globalization, and the resulting incapacity of national state regulation,
- increasing risk awareness, reflectivity, and self-reflexivity among consumers,
- a general move from product- to consumer-orientation in all areas of policymaking,
- an increasing hegemony of neoliberalism, including deregulation, privatization, and a high prominence of free-trade values.

In the case of organic food, KRAV started as a social movement initiative with the Swedish Ecological Farmers as a key driver. During its growth and increased public recognition, more resourceful actors, such as retailers and processing industries, advanced their positions and they are increasingly the ones who acknowledge the demand and suggest new types of goods that should be labelled. In marine certification, Unilever played a proactive role and cooperated with WWF in establishing the MSC. In forest certification, DIY chains such as IKEA, B&Q, and HomeDepot played proactive roles. Some of them appeared to be more reactive in their behaviour after being targeted by the direct action of EMOs. Yet, the retailers played central roles in their promotion of certification and – indirectly – in labelling. Another example is the European Eco-labelling Board, stating the mutual benefits of 'active cooperation of the EU Eco-label with retailers' as 'strategic partners'.[91]

Retailers may embody several types of encouraging arguments (see Chapter 6). They identify positive economic values in green niches or in green PR-making at the same time as they see green labelling as a means to reduce the risk of negative publicity. Being active in green labelling can therefore be part of a market-maintaining strategy. Retailers are often big and highly visible market players, and are therefore good targets for social movement campaigning. Retailers have direct links to end consumers and want to reach the concerned consumers who have strong political identities. Retailers provide 'quality choices' to consumers in contexts of individualization, life politics and identity politics (Marsden et al., 2000). Failure to do so entails a clear risk of losing competitive market space. Retailers project themselves as socially progressive, as part of the solution, rather than as part of the problem.

In a market-based governance system it is essential for the standard-setter to include business actors with 'structural power' (Clapp, 2005, p. 287), that is, with central network positions that can address powerful demands along the entire chain of production and distribution (Green et al., 2000; Cashore et al., 2004; Gulbrandsen, 2004; Jordan et al., 2004). Big retailers are such actors. Their involvement is crucial both in coalitions and in the labelling organization. They often have unique knowledge about feasibility, market potentials, consumer behaviour and concerns, as well as technological options. Retailers are even seen as representatives of consumers because of their structural position and the lack of strong consumer organizations (strong in terms of both resources and knowledge). 'The "reflexive turn" by consumers, and the individuation and identity politics involved in the consumption process ... tend to reinforce the political power of corporate retailers in their ability to "represent the consumer"' (Marsden et al., 2000, p. 79).

Although we have found their expertise and progressive stance generally appreciated among our informants representing other interest groups, both producers and social movements, several of them are at the same time worried about what they perceive as the increasing power of retailers. Retailers have forcefully issued their own 'green' trademarks such as Whole Foods Market, with its own certification scheme, and Coop's Änglamark in competition with both suppliers' trademarks and eco-labels. They present these trademarks as green labels although a possible criterion is that they could be based on existing eco-labels. A few informants express a fear that such trademarks will drive eco-labels out of business.

These players are also powerful actors in negotiations about labelling criteria. Representatives of retailers generally lean towards market pragmatism regarding the design of labelling programmes and the formulation of labelling principles and criteria. For example, they typically speak warmly in favour of free trade, of avoiding trade barriers in labelling criteria such as limits for transportation. Moreover, they do not want to alter their existing distribution channels, a lack of flexibility which often entails lengthy and seemingly irrational transportation of goods, such as production in one place, packaging in another place, and consumption back at the first place. A further issue is that food retailers stress that food products must look 'normal'; as we saw in earlier chapters, this is a framing that may conflict with values of naturalness.

The strong role of retailers has implications for the symbolic differentiation of green labelling in general, as they tend to prefer the market for certified and labelled products not to be too small. Although they generally acknowledge the need for (symbolic) differentiation – framed as facilitating environmentally reflective consumer choice – labelled products cannot constitute a small niche, they argue. In that case it would not be profitable to invest in it. In other cases, where certification and labelling are less a strategy to inform consumers and give them good buying options and more a strategy to avoid negative publicity and to secure the supply of sustainable raw material, differentiation is even counter-productive for their strategies. For example, in the forest sector, DIY retailers are generally merely concerned about ensuring sustainable raw material, and certified raw material would provide such assurance (Cashore et al., 2004). The retailers' position would be that the more available certified raw material it is possible to find, the better it is.

In sum, the inclusion of powerful retailers gives significant market-penetrating power in green labelling. The increasing involvement of retailers does not necessarily jeopardize the ideal of 'symbolic differentiation', of distinguishing green from conventional products, since they typically want to market certain high-quality products. Yet, it is clear that a trend towards increasing prominence of retailers at the same time results in a trend towards mainstreaming and market pragmatism away from stringent EMO values in green labelling.

**Producers: big and small**

In one way or another, green labelling requires the involvement of producers. Producers make the goods that are to be labelled. Alternatively, they supply certified raw material to other producers and retailers so that they can label products.

As regards mutual funds, it is not obvious who the 'producer' is. If the funds themselves are defined as the products (as they often are in the mutual fund industry), the banks or other providers of mutual funds are the producers. In our studied cases, it is the banks and the other providers of mutual funds that have the final word in criteria-setting for the ethical and green mutual funds. If the 'physical' products are in focus, the producers are plausibly the companies producing these goods or services. Criteria-setting and screening of green and ethical mutual funds obviously involve the producers of mutual funds, but also the producers of goods and services. For instance, as Lewis and Mackenzie point out, 'it is routine for most fund managers (whether ethical or non-ethical) to research companies and meet with them' (2000, p. 217). On the websites of banks and other funding institutions in many countries it is typical to provide a list of companies with which the fund managers have regular meetings. The producers of mutual funds often point out the pressure that they exert on the companies to modify the production and services in environmentally and ethically benign ways: '[W]e have engaged numerous corporations in discussions on a wide range of issues, from sweatshops to the environment' (Domini, 2007).[92] To be sure, far from all the dialogues with companies could be categorized as attempts at putting green and ethical pressure on them. In the United Kingdom, for example, Lewis and Mackenzie maintain that:

> It is likely that for the majority of ethical funds in the U.K., engagement is limited to questions from the fund asking for clarification on company policy, and informing the company of the fund's ethical policy. This activity is unlikely to do much to persuade companies to change their policies. (Lewis & Mackenzie, 2000)

Still, as the mutual funds are getting larger and more influential it is reasonable to assume that the potential to influence companies is increasing. A few years ago, Lewis & Mackenzie noted interesting examples of progressive mutual funds in the UK putting pressure on companies.

---

*Example of mutual funds trying to influence companies*

Three ethical fund managers in the U.K. do have policies of engaging with companies in order to persuade them to change their policy, namely Friends Provident Stewardship, NPI Global Care, and Jupiter Ecology. Friends Provident Stewardship has £900 million of funds under management. Jupiter Ecology and NPI Global Care have over £100 million between them, so these funds account for over 60% of the ethical fund market. These funds have used a number of different procedures to pursue engagement, including writing letters, holding meetings with managers, doing sector surveys and feeding back the results to management, as well as attempting to lead policy in more general ways, by writing articles, briefing the press, giving addresses at conferences, participating in industry wide initiatives. (Lewis & Mackenzie, 2000)

---

A far less explored question than the influence of mutual funds on companies is to what extent pressure is exerted in the reverse way, namely from companies to mutual funds. In principle, such pressure could either be exerted through calls to raise or lower the ethical and environmental bar within mutual funds, depending on the current level of ethical and environmental efforts in the companies at stake. The extent to which companies have tried to lower the environmental bar of mutual funds is, of course, something that neither party is keen on speaking about aloud. To learn more about it – which would be at least as interesting as in the other cases where we have better data regarding backstage processes – would require comprehensive and in-depth exploration of the interaction between mutual funds actors and companies. For now, we can mention that there is a mutual interest between both parties to collaborate. Companies have an interest in increasing their symbolic capital (as well as pecuniary capital by getting access to investment capital); the SRI industry has an interest in increasing the range of companies to invest in, and often gaining an eco-pragmatic (or ethico-pragmatic) image: neither ruthlessly exploiting nor compromising in terms of investors' profit. Thus it is only reasonable to assume that the two parties exert mutual influence.

Although the producers within the financial sector are key drivers in green labelling, we see a common pattern that actors at the beginning of the production chain are less enthusiastic than actors in the middle of the chain. Often, producers are not very eager to label or certify their products; they develop a wide arsenal of objections. It is common that big retailers or big processing industries enforce suppliers to seek certification (e.g., Marsden et al., 2000). Forest companies, for example, have clearly been less eager to join the FSC than have the retailers, but forest companies have felt more or less compelled to do so (Domask, 2003) and the case of marine certification reveals the same pattern (Constance & Bonanno, 2000).

One main reason why producers are less eager to adopt standards is that they have to bear most of the costs because of following the standard criteria. It is producers, rather than retailers, who have to make on-the-ground changes, and who have to be audited by a third party. In general, standards in labelling focus more on production and less on distribution and transportation. This is true even in cases where there is an expressed ambition to certify the entire production chain, such as in the seafood case. Companies at the beginning of the production chain have, moreover, a greater distance from end consumers, and might not experience any consumer pressures. Big producers, however – similarly to big retailers – can be highly visible, which makes them a good target for protests by social movement campaigning.

Companies that produce on a small scale may therefore be even less willing to bear the costs of complying with green labelling criteria. There are many reasons to believe that it is easier for big companies than small companies to adopt green labelling. Small corporations usually do not have the same capacity or financial resources to comply with standard criteria. Nor do they have the same resources to pay for audits. Although several schemes have criteria that discriminate positively for small business (e.g., by paying lower or no fees for having their products labelled, as with the USDA's Organic Seal in the

United States), it is generally easier for big business to use green labelling schemes. It can be easier for big companies than small ones to adopt standards, simply due to economies of scale. As to organic food production, it has a history of small-scale thinking and ideology; but in practice organic farms are on average larger than conventional ones (Guthman, 2004; Rydén, 2005). The FSC has invested much energy and prestige in social equity issues in its own democratic and inclusive organizational structure and in its standards (such as recognizing small-scale community-based forests in developing countries). Thus, it is perhaps a painful experience that mostly big companies from Northern countries are FSC-certified. Thornber maintains:

> Forest enterprises which are not familiar with formal, documented management systems and concepts of inspection, but which nevertheless produce sustainable results through less formal checks and balances, are likely to be at a disadvantage. (2003, p. 68)

Large and multinational companies are more likely to have the technical capacity, management structures and skills to implement standard requirements and to market their products. Small enterprises (often family-owned or community-based) have less technical and financial flexibility, and may be more averse to risk. They do not have the same financial robustness to bear the costs and risks. It is also common for large companies to enjoy superior access to information about existing and prospective certification requirements. When commenting on the forest case, Meidinger (1999) concludes that the stricter the environmental standards become, the greater the disparity becomes between the abilities of large, sophisticated corporations and small, often third-world businesses to meet them.

An additional reason why non-industrial forest owners, farmers or fishermen are often less willing to support certification and labelling may be that they value independence more highly. In our cases, we see that associations representing farmers, non-industrial forest owners, and fishermen typically stress the value of independence. Representatives of these associations claim that their professions have long been advocates of sustainable practices. Early on, they argue, they followed responsible ways of preserving the natural resources upon which their business is dependent (see also Törnqvist, 1995; Cashore et al., 2004, p. 202). They question the need for such external actors as EMOs and retailers to set standards that they are compelled to follow, as they already feel overburdened by existing rules.

However, we should not exaggerate the big-versus-small split, as we also see evidence of small actors going in for labelling. There are sometimes disadvantages for big companies in adopting green labelling. For example, big technological investment may create such inertia that it is difficult to undertake the kind of continuous revisions that labelling programmes often require (see Chapter 7).

## Environmental movement organizations and other social movement players

EMOs are often the kind of actors that stage consumer boycotts, provide consumer recommendations, and promote consumer reflections. They also often take initiatives in building coalitions for labelling as well as establishing labelling organizations. Even if labelling is not initiated by EMOs, most initiators understand that some EMOs need to participate in green labelling, at least as advisors. The participation of generally appreciated EMOs, animal-rights groups, and other SMOs can bring the necessary credibility to the project because of their 'moral authority' (Hall & Biersteker, 2002), for example, reflecting collectively shared values, voluntarism, and consumer power (cf. Boli & Thomas, 1999). Such actors can also be sources of new framings and green expertise (Boström, 2003a, 2004a). Such EMOs as WWF, FoE, Greenpeace, SSNC, and Sierra Club help to clarify, concretize, and popularize the meaning of commonly held ideas about natural food, animal protection, precaution, sustainability, and biodiversity. Moreover, labelling organizations tend to see substantial individual membership-based EMOs as reflecting strong consumer power.

From our cases of seafood labelling, forest certification, organic labelling, green electricity, GM food, and paper we see that EMOs and other SMOs are deeply engaged in the early agenda-setting phases of labelling processes. In the SMO-governed labelling (e.g., Good Environmental Choice) an SMO has sustained power by definition. Since labelling is often organized in hybrid ways, however, it is interesting to ask whether and how EMOs sustain their power in such organizations.

It is clear that such a big EMO as WWF is continuously involved in several sectors in providing constructive viewpoints, participating in standards revisions, organizing 'groups of buyers', and thereby stimulating the demand for labelled products. Yet, during the early establishment of organic food labelling in the United States and Sweden, for instance, the role of SMOs was more pronounced than in its administration and further development. Environmental, animal-rights, consumer, and organic movements tend to take fewer new initiatives within the established labelling organizations and instead act in a reactive fashion. In KRAV, they participate in discussions when the issues fit their current priorities. In other words, they work as watchdogs, ensuring their priorities are adequately dealt with and covered.

Always acting proactively is difficult. A critical point is that SMOs simply have fewer resources than the industrial players have for participating in routine activities. As seen in other studies of standard setting, member groups do not have the same opportunities to participate, simply because of their unequal access to financial resources (Schmidt & Werle, 1998; Cochoy, 2004; Tamm Hallström, 2004). Labelling organizations themselves have a scarcity of resources, and cannot assist weak actors to any larger extent, in order to help them contribute regularly. Labelling organizations have to rely on existing organizations' resources.

Moreover, SMOs must prioritize. Their priorities are not only based on financial considerations, but also reflect their own collective identities (Boström, 2004a, 2006b). Members of SMOs may think that participation in labelling activities is not a top priority. They must deal with other topics and strategies as well, and must therefore be conscious of their (radical, progressive) movement identity. Indeed, there is something to the view that movements continuously aim at opening up new conceptual spaces (Eyerman & Jamison, 1991), thus avoiding institutionalization. The role of social movements is more about '*making* politics' than about '*administrating* politics'.

Nonetheless, SMOs' occupation of board and committee positions in labelling organizations is of significant value. SMOs have latent power resources in the option of officially ending their membership or participation in the labelling organization, or refusing to take part (such as in PEFC and SFI), which would hurt the legitimacy and credibility of the arrangement.

SMOs have to consider the risk of becoming co-opted by business interests due to their engagement in labelling; they have to make sure they preserve their cognitive and organizational autonomy (cf. Boström, 2004a; Boström & Klintman, 2006a). One difficulty concerns the ways in which EMOs should create a balance between preserving critical distance and at the same time remaining loyal to their cooperating partners in the labelling organization. How should they act if and when they observe that a certified company does not act in accordance with the standard? What behaviour is expected from the EMOs? While they can make an impact as members or participants in the arrangement, they can also play effective roles as outsiders, for instance by watching how certified companies comply with labelling criteria. Alternatively, they can insist on other measures being taken that are not covered by labelling criteria. Deep involvement in labelling organizations may hamper such external roles because certified companies may expect cooperating SMOs not to stage official protest campaigns against the labelling organizations and certified companies. The SMOs themselves may also have an interest in not tainting the reputation of the labelling organization, since the SMOs have invested time, resources, and prestige in it. For example, WWF has expressed strong commitment to such labelling organizations as the FSC and MSC. It would probably hurt the name of WWF itself if these organizations were severely attacked.

However, there is not necessarily a trade-off between insider and outsider strategies. SMOs that participate in labelling projects are empowered in various means through their participation. They gain important insights about labelling visions, goals, concerns, principles, policies, interpretations, and so on. They can have great specific expertise about the labelling practice, which they may use to assess actual performance. In labelling schemes, observations of failures to comply with standard criteria should engage procedures that may lead to sanctions in the form of withdrawals of a certificate. SMOs can address concerns directly to standard-setting or certification bodies or remind corporations to follow the standards properly. Furthermore, SMOs can continue mobilizing coalitions around a labelling organization even after it has been established.[93]

## Consumer representation and participation

Elsewhere in this book we have claimed that labelling can be perceived as part of a broadening of power in society, which is beneficial to democracy. A relevant question regarding participation and political consumerism is the following one: Must a decision-making process involve active involvement of unorganized consumers to be classified as an example of consumer democracy? How about deliberation across organizations at the meso level? As the reader has seen, this book takes the stance that the *participation of organizations* could very well be analysed through the lens of political consumerism. Still, additional questions ought to be raised about whether the organizations are inclusive or exclusive with regard to consumer groups, input from 'the public', and ideas from outside, *in the development* of the consumer-related policies and criteria. To assess to what extent organizations and sectors involve democratic participation of the public, we need to examine the degree of public inclusion within the organizations. Whereas the monetary side of political consumerism (mainly the purchasing of goods and services) makes citizen-consumers unequally powerful (due to their unequal monetary resources), the supply side where tools for political consumerism are developed might be more inclusive, at least in principle. In Chapter 8 we mentioned the Big Three debate, which illustrates the most extensive consumer participation in labelling debates so far, to our knowledge.

The Big Three, however, is more of an exception to the rule that voices of consumers are represented through organizations. In Sweden, consumer associations are comparatively small (which has a lot to do with a strong consumer agency). One informant from KRAV mentioned that they often did not even contact a particular consumer association that is a member of KRAV, in various discussions regarding labelling criteria, because they know from experience that this organization seldom participates because of lack of time and resources. In their eyes, big membership-based EMOs such as SSNC or big retailers such as Coop (with three million 'members') better reflect consumer power and voices. Even KRAV itself, although it is a hybrid organization reflecting different interests, is sometimes seen as representing the consumer view. To be sure, KRAV has lots of direct experience with consumers who contact KRAV directly, and it uses various techniques for gathering relevant information. In the seafood labelling case, KRAV conducted a survey to obtain input about consumer priorities. Ironically, what appeared as the top consumer issue – that labelled seafood should not include toxins and heavy metals – appeared to be the issue that was most tricky to standardize with stringent criteria (Lönn, 2003).

KRAV, like other labelling organizations, represents – directly or indirectly – large parts of the public, as citizens and consumers, at the level of principle. However, there are large variations on a scale between active and passive membership. SSNC has a democratic structure with extensive individual membership, so in principle each individual's view is represented. Yet, the issue of thresholds and criteria for eco-labelled electricity is something that is treated and decided through internal discussions and deliberations, mainly between SSNC staff, the authorities, and energy companies. The same is largely true in issues of labelling of forestry and food: the most active and professional actors in the organizations primarily take care of such decisions, tacitly supported by the passive members. Furthermore, several informants from various sectors hold that the other strong stakeholders (aside from large NGOs), which are the most active driving force for introducing and setting labelling and its criteria, are indeed the representatives of businesses (see also Gulbrandsen, 2006; Holzer, 2006). In the case of forestry, an informant from a forest company claims that it is not 'the man in the street' who plays any major part, it is rather the consumer-focused businesses that make demands. The business has the interest in obtaining a quality seal for the product.

This said, how can informal groups and individuals in the general public make their voices heard in ways that can be called participatory? Although the 'unorganized' public are not said to have been the first movers of green labelling in any of our cases, it is almost a truism to say that the labelling schemes are dependent on whether the public in their role as consumers accept and hold a certain trust in the labels. In addition to this important 'monetary political consumerism', the more 'discursive political consumerism' on the supply side among the public can also be found outside the organizations and the most powerful stakeholders. Debates and inputs are often broader than the intra-organizational activities, and labelling schemes are sometimes affected by the broader public debates. Again, the Big Three is the most obvious example.

## Conclusion

We hope this chapter has shown that the organizing process is an essential factor in the establishment of a labelling programme. Among other things, the organizing process can have outcomes in terms of

- how standards are formulated;
- whether competing standards are established;
- how political debates and frames are developed;
- whether a reflective trust among stakeholders develops (see next chapter);
- whether power shifts occur through time; and
- whether stalemate in negotiations and decision-making is common.

Organizational issues are themselves hot topics for debates among the stakeholders. Perhaps debaters in the arrangements sometimes even put too much emphasis on organizational form, because it is not the case that form determines standard content, as we have stressed in the chapter. For instance, if leaders of a labelling programme have a sufficiently open and responsive mind, EMOs can play critical roles also as advisors or outsiders, and consumer voices may be taken into account. Another aspect of the SMO–business interaction concerns the mutual mistrust among such groups. We now turn to this issue.

# 10
# Dealing with Mutual Mistrust

## Introduction

In dealing with trust relationships, there are two types of challenges that labelling initiators have to confront. The first one concerns the relationship between consumers and the programme/scheme (see Chapter 4). The second one, and the topic of this chapter, concerns the relationship between the actors participating in the process of standards development. Various groups – including consumer groups – that are seen as important to include as participants in labelling programmes may mistrust each other, even hate each other. To develop standard criteria may require that groups representing, for instance, social movements and business can initiate a dialogue and reach agreements. However, they may lack experience of such dialogues across societal spheres, and may not be very keen on communicating on equal grounds. They may doubt the other stakeholders' experience and competence, and may believe that the others are pushing hidden agendas.

In this chapter, we investigate why, and how, groups in some of our cases have managed to develop a kind of mutual and reflective trust, in spite of their initial mistrust. We will also analyse why such rapprochement has failed in other cases. In doing so, we analyse the role of repeated interaction and cognitive authorities, as well as transparency and auditability, terms that we explain below.

## To trust or to mistrust:
## the role of repeated interaction

If X trusts Y, X believes that Y has the right intentions towards X, and that Y is competent to do what X trusts Y to do (cf. Hardin, 2006, p 17).

Focusing on intentions and competence is appropriate with regard to our empirical findings. Several informants representing specific stakeholders in our cases mention how lack of trust in the intentions and the competence of other stakeholders made them hesitate before getting into labelling. Many of them suspected hidden motives and purposes among other parties. For example, informants representing forest industries mentioned their suspicions regarding EMOs' eventually hidden agendas. This suspicion made them reluctant to join cooperation within the FSC: 'Won't they just require more and more of us as soon as we reach initial agreements?' Some of them referred to previous negative EMO reporting and campaigning as a reason for their hesitant attitude. EMOs for their part expressed concern that business actors may participate in labelling programmes and contribute to the implementation, but that business thereafter does little to perform properly and to comply with the standards. There were furthermore numerous comments in our interviews about other actors' limited competence and experience. For example, fishermen mentioned what they perceived as factual errors in a seafood consumer guide recently issued by the WWF, a list that included blacklisting of some seafood products sold on the market. This was seen as a basis for questioning the competence of this organization.

Trust is a relational concept (e.g., Warren, 1999). Trust develops from repeated interaction. It is constituted by knowledge, experiences, perceptions, and beliefs about the trustworthiness of the other. Actors gradually learn that interacting actors are trustworthy, but such learning requires communication and interaction. Trust always involves risks. One can never be sure that interacting actors are truly trustworthy, although they might appear so. Often it is easier to mistrust than to trust someone. Hardin suggests there is an 'epistemological asymmetry, between trust and distrust' (Hardin, 2006, p. 18). Trust requires a certain kind of knowledge about actors and institutions that is difficult to acquire, whereas we can *dis*trust someone or something almost without any knowledge. As a rule, according to Hardin (ibid.), 'we trust only those with whom we have a rich enough relationship to judge them trustworthy, and even then we trust only over certain ranges of actions'. In our view, however, Hardin is partly wrong here. A good part of the public may well hold a 'simple', excessive (blind) trust in various institutions (such as green labelling schemes). This is likely in political cultures such as those in Northern Europe where people tend to trust abstract institutions more than people in other regions do (cf. Kjærnes et al., 2007). Yet, as regards the relationship between such stakeholders as social movements and industries, one may speak of an epistemological asymmetry, which favours mistrust. Earlier experiences with a confrontational environmental movement can be a firm basis for mistrust, as we noted above, and it might require much interaction before people representing such different societal spheres are willing to cooperate and even talk to each other. In the Swedish seafood labelling case, fishermen were not even willing to sit in the same room as environmental activists. How could they initiate a dialogue – let alone cooperate – in these circumstances?

We have found strong evidence in our cases that repeated interaction in inclusive organizational settings has fostered mutual (reflective) trust among participants. Groups have gradually learned that they can trust the intentions and competences of other parties in the arrangement.[94] One fisherman involved in the seafood labelling case said that the group he represents was surprised at KRAV's responsive attitude. Fishermen also gradually understood that other actors, for instance EMOs, involved in the project had no hidden agenda aimed at destroying the fishermen's business. Participating in KRAV's project for the very first time also showed the fishermen that there were (private) rule-making authorities that were really interested in understanding their fishing practices. People from the project group at KRAV joined fishing trips to learn about their practice so that KRAV would be able to develop rules that could work in practice. One fisherman whom we spoke to had never heard of officials from fishing authorities joining fishing trips, and expressing such interest and curiosity in the reality of the fishery profession. On the contrary, he had experienced only traditional rule-making authorities imposing strong and capricious rules and regulations without asking fishermen about their views and knowledge. It is very clear that the responsive attitude among the project leaders in KRAV fostered a reflective trust among the fishermen. The latter group also gradually understood that organizations such as WWF did not really want to undermine the fishery business. One fisherman summarizes his experiences:

> During this process, we have understood others' arguments better. We have better understood what they are aiming for. We have always seen us threatened because we have seen that others want us away. But now we see an opportunity. We haven't been sitting together before.

However, it requires management skills and plenty of effort to start such a positive spiral of trust development. Even if a dialogue is initiated, it might be unsuccessful, and it might only amplify mutual mistrust. In the US controversy over whether it should be allowed to label GM food with the USDA's organic seal, we have found several examples of mistrust – deliberate or otherwise. The American Crop Protection Association [ACPA], which favoured the inclusion of GMs, mistrusted both consumers and consumer groups that had called for exclusion of GMs from the organic label:

> We also question whether consumer expectation or consumer opposition are valid or appropriate reasons for determining whether a production method is compatible with organic production. (ACPA, 2000, p. 4, in Klintman, 2006, p. 433)

The ACPA misunderstood that the consumer opposition was based on the consumers' views of the intrinsic value of consumer power, the right of consumers to decide for themselves, aside from the scientific aspects and risk levels of GMs (Klintman, 2006).

As we all know from daily life, experiences of not being listened to, or of difficulties in communicating one's viewpoints, can trigger mistrust. Informants representing paper producers, and those who promoted Paper Profile as an alternative to paper labelling, mentioned poor support and communication within the Nordic Swan as one factor behind their mistrusting attitude (Nilsson, 2005). The paper producers were not involved in the Swan – they were only represented very indirectly through their 'meta-meta-organization', the Confederation of Swedish Enterprises. As they saw it, they were thus insufficiently represented.

Even in the cases where reflected trust is established, it is only the actors included that hold, and are subject to, this reflected trust. A huge group that is outside the labelling arrangements is the consumers. Also, increasing trust among some actors may be supplemented by increasing mistrust among others, that is, among other actors who do not take part in the labelling arrangement, but are indirectly affected by it. We illustrate this with the following example.

---

*Example: Symbolic differentiation and mutual mistrust* An unfortunate development in the Swedish forest case is the clashes between the environmental movement and small forest owners, where the latter 'group' joined the FSC competitor PEFC. EMOs speak in favour of the FSC and criticize the competing programmes, because they see that the FSC has stricter standards. 'So why not use the stricter standard?', EMOs argue. However, as small forest owners insist on following the competing model, the unintended consequence is that EMOs help big Swedish, FSC-certified, forest companies in their shaping of a green image, while implicitly targeting small forest owners as the bad guys. One could say that the associations of private forest owners in Sweden have themselves to blame for not committing to the 'most stringent' standard. However, the fact that small forest owners typically have a negative attitude to detailed regulation and abstract standards should not be taken as evidence that they are uninterested in environmental protection (Törnqvist, 1995). It is difficult to ignore the fact that a symbolic differentiation has been created that favours big companies, and which is bracketing their long history of environmentally destructive forest practices. Hence, the old saying 'big is beautiful' gets new wind in its sails, whereas the fact that also small, environmentally conscious forest owners might be able to do several good things for the environment is overlooked.

The broad factors that we have analysed so far in the book – framing, organizing, and policy context – are, we argue, critical for understanding how stakeholders learn to trust each other in a reflective way. For instance, an inclusive organizational setting – with many hearings and meetings and many opportunities for face-to-face interaction and for sharing viewpoints – can facilitate the development of mutual reflective trust. Such a process requires time, as several informants underline. Trust does not develop overnight. In some instances, a conscious framing process, including frame bridging and frame reflection, can facilitate the development of mutual trust. Through such framing processes, antagonist actors can develop a degree of mutual understanding of each other's ideas, intentions, concerns, and arguments. Once participants develop conceptual links between environmental themes (e.g., natural products) and economic reasoning (i.e., natural products must also be tradable and profitable on a market), a discursive platform for developing mutual reflective trust is established. Furthermore, a fairly consensual political culture helps actors to trust that it is possible to reach compromises, despite deep initial controversies.

Through various organizational means and framing devices, in addition to those mentioned, labelling organizations themselves may strive – explicitly or implicitly, consciously or unconsciously – to create mutual reflective trust among stakeholders (and towards the general public). In the following sections, we will analyse the role of science and cognitive authorities as well as transparency and auditability in such endeavours.

## Science and cognitive authorities

Since green labels are categorical claims – implying that labelled products are better for the environment, for health, for animal welfare, for social justice, and so forth, compared with competing 'conventional' products – an immediate need emerges to establish credibility and legitimacy with reference to authoritative knowledge claims. Whereas science alone would be insufficient as a cognitive base for rule-setting, it would still be impossible for green labellers to ignore science. Authoritative knowledge statements are typically provided by science.

To be sure, many scholars contend that science in late-modern life is contested. The scientific community faces difficulties in maintaining cognitive authority, or at least in maintaining a monopoly position. Interest groups increasingly cast doubt on the value, 'objectivity', or 'disinterest' of science and scientific research. Uncertainty, complex causal relations, unpredictability, disagreements among scientists, and vested interests are factors that challenge the hegemony of science (e.g., Beck, 1992; Yearley, 2005).

At the same time, science is sought after in an increasing number of societal sectors and areas. Another aspect of this 'spreading' is relevant to the theme of this book: more and more people are scientifically educated. As a consequence they learn to criticize science and conduct their own research within political organizations, business contexts and civil society organizations. Knowledge producers operate in an organizational context. In this way, the success of science and scientific reasoning makes it more difficult to uphold strong demarcation between science and non-science (Nowotny et al., 2001).

Given these tendencies, attempts at defining a 'purely and neutrally scientific' basis for environmental standards and criteria, from which disagreeing and mutually mistrusting groups can develop common views, may therefore be a poor and shaky exercise. Yet, somewhat paradoxically, scientific reasoning appears to permeate politics and society to an increasing extent. Scientific validity and scientific consensus are collectively shared ideals in all types of policymaking, but particularly in environmental policymaking (cf. Lidskog & Sundqvist, 2004), and can therefore be instrumental in the rapprochement of mutually mistrusting groups. The fact that science spreads into society and loses its monopoly in determining truth in no way leads to its demise (cf. Collins & Evans, 2002; Jasanoff, 2005).

As to labelling, we note that labellers very often try to convince various audiences that their labelling criteria are based on relevant, adequate, valid, robust knowledge – preferably scientific knowledge. Likewise, science and cognitive authorities play important roles in the debate and for the trust relations among stakeholders in our cases; but there are also great differences across the cases as to how they play a role (Boström & Klintman, 2006b), as the following discussion shows.

In the American GM labelling controversies, scientific understandings have not led to any mutual trust of the parties holding opposing views. In Chapter 8 (the section on frame resolutions), we noted that the issue of whether or not GM food should be allowed to use the organic label was not resolved by a common scientific understanding. To be sure, both sides tried to make strong scientific claims. The groups in favour of such an inclusion of GMs referred to the scientifically based 'substantial equivalence', whereas the opponents of inclusion – albeit often rhetorically – appealed to a very strong trust in the 'science' of labelling.

Instead of a science-based resolution, that controversy was resolved through a reframing into consumer democracy: Regardless of scientific claims about risks and safety, consumers should have the right to choose 'GM-free' food (see also Klintman, 2006).

In the other GM labelling controversy in the United States – about whether GM food should have a mandatory label – the antagonistic coalitions have used scientific claims in an interesting way. Due to the fear of negative associations with GM food, the coalition in favour of GM food typically puts up vigorous opposition to a mandatory GM food label. This is obvious enough. What is interesting is that they have a very strong faith in science when they discuss the GM technology as such, its safety, the objectivity of scientific results and so forth. However, when they discuss a GM food *label* they move from being scientific optimists to being gloomy scientific pessimists, by sounding like the most postmodern judgemental relativist. Thus, they reject this labelling by claiming that no statements can be made about a label that are not highly dependent on social and political processes, and the financial interests of the 'Anti-GM alliance', and of the human errors involved in controlling the labelling system, and so forth.

The proponents of such a label – who are typically GM opponents – use the same arguments, but in reverse. They act as judgemental relativists – highly critical of science – when they discuss the GM technology as such, by claiming that the 'science' behind the technology is highly uncertain, entirely dependent on the financial and political interests of the GM industry with whom the US government cooperates. When discussing GM food *labelling* the GM opponents express a very strong scientific faith, almost an objectivist view that a mandatory GM label, its thresholds, and control system would provide 'purely scientific' knowledge: the perspective of epistemic absolutism.

We see a parallel with the 'boundary work' between science and non-science that Gieryn (1983) talks about. In the cases mentioned above, however, it is not primarily the researchers themselves who try to construct boundaries against other intellectual practices. Instead, it is policymakers, rule-makers, SMOs, and other stakeholders who try to legitimize their goals and strategies by giving these a scientific frame. In the case of mandatory GM labelling in the United States, these efforts have led to what Klintman (2002a, b) calls 'argumentative *crossovers'*, defined as 'cases where an alliance uses a certain type of argumentation for one issue [e.g., GM science and technology], but shifts to using an opposing alliance's type of argumentation for a closely related issue with different practical implications [e.g., GM labelling]'. At one level the extensive polarization (including the crossovers) between the GM-supportive and GM-sceptical coalitions in the labelling issue reflects a high degree of internal efficiency (for the respective coalition). We have seen, however, how such argumentative flexibility works against the goal of mutual, reflective trust, both across the coalitions and *vis-à-vis* the public. When arguments within an coalition appear to be self-contradictory, the cultural resonance and public support are likely to be low (Benford & Snow, 2000). Moreover, GM opponents' treatment of mandatory GM labelling as reflections of an absolute truth may be effective at an early stage, by creating a public opinion in favour of labelling. Indeed, a vast majority of the US public is in favour of a mandatory GM label. However, to maintain such public trust, and to gain mutual, reflective trust *vis-à-vis* policymakers, would most likely require an epistemically conscious and frame-reflective debate about how to handle knowledge uncertainties of labelling (cf. Klintman, 2002a, b).

In contrast, in our Swedish forest case, science – in a broad sense – has constituted an important platform for discussing and negotiating labelling criteria. People representing the forestry and the EMOs were not too convinced of each other's knowledge on the issue. Yet, both groups agreed that Swedish natural science within the forest sector showed good records and that natural science should play an essential role in discussions about criteria. Both groups expressed strong trust in science. Informants from both parties maintained that it was crucial to refer to existing scientific research during the whole standard-setting process. For example, an official report (SOU, 1997/97), written by recognized Swedish researchers, played an extremely important role. It states the protection measures needed and it assesses the percentages of forest areas that should be protected from exploitation in order to save biodiversity. This official, science-based report was an important document for EMOs to refer to when they exercised pressure to bring about strong protection measures in the FSC standard. While conceding to these pressures, the forest business systematically downplayed other calls for prescriptive, substantive standard criteria in areas where they had found scientific evidence of environmental improvement less clear. This concerned, for example, the consequences of introducing exotic species after having felled parts of forests. It also concerned preventing the use of chemical fertilizers and pesticides in forestry. Although EMOs disagreed with the foresters' substantive claims on these issues, EMOs agreed with the principle of giving science – and the notion of 'scientific evidence' – a prominent role. The parties could agree by referring to 'further research' and 'further revision of standard criteria' so that possible new evidence would become the basis for solving the issue in the future. This shows a firm trust in natural science and research, a trust that itself reflects the strong role of natural science in Swedish forestry historically (Törnqvist, 1995), and even in Swedish political culture in general (Jamison et al., 1990).

In the seafood case too, science functioned as a legitimate platform for the introduction of new rules. It was a basis for compromising and consensus-making. It is interesting to note that actors succeeded in finding common ground by relying on cognitive authorities, despite their initial mutual mistrust (Boström, 2006a). Fishermen used to criticize the knowledge that researchers produce, and the recommendations resulting from their knowledge. The researchers, in turn, generally ignored fishermen's experiences, for example, of changes in the levels of fish stocks (Hasselberg, 1997; Johansson, 2003). Researchers had argued that fish stocks are declining whereas some fishermen had argued that the fish move depending on such factors as water temperature shifts. Knowledge uncertainty left room for disputes and made it hard to create a dialogue between the parties.

Nevertheless, the project found one important way to handle this mutual mistrust. The International Council for the Exploration of the Sea (ICES) was an important reference point – a cognitive authority that significantly helped groups agree on basic premises regarding assessments of the status of fish stock and catch options. ICES, founded in 1902, is an IGO that coordinates scientific research. It gives advice on fisheries management in the North-East Atlantic. The cognitive authority of the ICES is generally recognized: 'ICES' influence on regional decision-making depends largely on the organization's reputation for scientific excellence and neutrality towards the often competing claims of various stakeholders – which is widely perceived as high' (Stokke & Coffey, 2004,

p. 119). Among other things, ICES has been recognized for further developing and operationalizing the precautionary approach in fisheries management.

ICES is not entirely uncontested, but all major stakeholders in the Swedish context recognize the importance of this organization. Therefore, all agreed that the labelling scheme should rely on ICES' assessments. This was a key factor explaining openness for dialogue and negotiation. Yet, debates continued and concerned how much the regulatory framework should be dependent on ICES' advice. Should the certification body approve only of stocks that ICES considers to be within safe biological limits, or is it sufficient if the certification body merely considers ICES' recommendations? One problem for the project was that ICES does not make assessments for all species and stocks relevant to eco-labelling consideration. The Project Group gradually came to an understanding that they have to refer to ICES in order to gain legitimacy, but the framework could not fully rely on ICES. Some stakeholders, such as the SSNC, were disappointed that ICES was not given an even more central role. Informants predict that tough subsequent debates and disagreements will continue, about stock assessments and the role of such cognitive authorities. Yet, it is clear at the same time that reference to ICES helped to establish a platform for dialogue, and a framework within which compromises could be made.

We do not find the same level of trust in science in the Swedish organic food case, but rather a more reflective and *ad hoc* use of scientific research. In contrast to forest and fish issues, which are agendas well-established within the environmental movement, labelling of organic products stems more from an independent movement: the organic movement. The organic movement developed from an explicit critique of conventional agriculture and its associated scientific foundation. This particular history brings up another view of the role of science in labelling. Scientific results appear in the Swedish organic case on a more *ad hoc* basis. It is fair to say that science in this case is more subordinated to ideology or firm principles concerning, for instance, 'naturalness'. Organic labellers are open to scientific results when such results fit existing understandings and framings, and reject research that does not fit, instead highlighting alternative scientific results that are more consistent with organic principles. For example, research on possible environmental (or health-related) benefits from the use of artificial chemicals in agriculture and food processing is deemed irrelevant at the outset; it is already defined in basic framings that the farming should be natural – without the use of artificial chemicals, that is.

However, if scientific results are available that support organic production, the organic network certainly refers to these results, or, if such 'evidence' is not available, they can nevertheless argue that some products are better than others, with explicit or implicit reference to the precautionary principle.

Perhaps it is possible to maintain this rather asymmetric, and in a way unsystematic,[95] attitude to science because we see comparatively little mutual mistrust among participating actors in this case (save for the initial years). In other words, there is less need for an external reference point such as 'scientific evidence' for the sake of reaching agreements among groups. While it is true that some researchers from the Swedish University of Agricultural Sciences (SLU) now and then speak out vehemently in criticism of organic production, a broad group of stakeholders, including not merely the environmental movement but such key (conventional) players as LRF, KF, and Swedish authorities, nevertheless defend organic principles and claim the environmental benefits. As long as they continue to do so, organic labellers could stick with their *ad hoc* – or reflective – relation to science.

What we can see from these sectors is that two views of science and knowledge compete in labelling processes. One is the reflective view that subordinates science to ideology. This is not exactly judgemental relativism implying that no knowledge claim – for instance through labelling criteria – is better than any other knowledge claim; however, it is closer to this epistemological pole than the opposite. The opposite view of science and knowledge relies strongly on the prominent role of natural science and the existence of 'scientific evidence'. We have often seen this 'epistemic absolutism' in various actors' calls for better, stricter labels that separate all bad substances from all good ones. To be sure, both these views of science and knowledge can exist and compete with each other within the same standard-setting process. Reflexivity, *ad hoc* use of science, and notions of uncertainty and precaution certainly materialize in the forest and seafood labelling cases as well. The more participating stakeholders mutually mistrust each other, the stronger role science as an external reference point can have.

Political culture also plays a part. David Vogel (2001) addresses the significantly greater impact of the precautionary principle in Europe than in the United States:

> [the] faith of many Americans in the capacity of risk assessment to objectively define a product or technology as 'safe' or 'unsafe' stands in sharp contrast to the situation in Europe, where the public embrace

of the precautionary principle appears to reflect a post-modern view of science, one in which scientific truth and thus risk assessment is socially constructed – and thus indeterminate. (Vogel, 2001, p. 20)

In sum, science and cognitive authorities play different roles in different labelling processes. Therefore, cognitive authorities may be more helpful in some areas, as external reference points for the development of mutual (reflective) trust, whereas science in other cases merely functions as a weapon in battles and helps only to prevent mutual mistrust. The role that science plays can, *firstly*, be connected to the traditional roles that science has played in the field where the regulation is located, and to whether there are any cognitive authorities there, whom several parties accept and trust. In the case of forestry, scientific knowledge, along with specific knowledge centres, has long since played a central role, entailing a strong belief in science in the labelling programme. In the fish case, we have already exemplified ICES as a strong cognitive authority. *Secondly*, the degree of conflict level in the field is significant, as we illustrated with the example of the US mandatory GM labelling controversy. *Thirdly*, social movements or other knowledge actors may play significant roles. Such actors, particularly if they are well-established, do not let themselves be overrun by new scientific evidence or recommendations. Organic farming is an example of such a sector. *Fourthly*, the more general political culture may certainly play an important part, such as whether the precautionary principle is acknowledged, and how it is framed and interpreted in the political culture.

## Transparency and auditability

A basic requirement for credible labelling is the possibility that someone external to the arrangement is able to ensure that the producer or seller of the labelled product really complies with labelling criteria. This external category could be the public, SMOs, an independent certification body, media, or any other actors. It is especially a concern among EMOs that vague standards and poor control would make it possible for a certified business to stick with business as usual. Two key standard-setting ideals of relevance for this discussion are transparency and auditability.

Criticisms of green and ethical mutual funds have largely concerned the *transparency* of the companies subject to a fund's 'green' or 'ethical' categorization, companies that have 'unsubstantiated and unverified social and environmental disclosures' (Laufer, 2003). By extension, there has also been criticism that mutual funds, despite their own calls for more transparency among the companies that they invest in, often compromise their own transparency:

> If ethical mutual funds expect full disclosure from the firms they are evaluating, it only seems just and appropriate that they engage in full and complete disclosure of their own activities. (Schwartz, 2003)

For example, when the magazine *Business Ethics* was to select the award-winning funds in its yearly Social Investing Awards, the judge was struck by the lack of disclosure of the funds, for instance concerning levels of charitable contributions:

> They declined to provide basic information that helps us judge their social performance. (*Business Ethics*, 2001)

The level of transparency among certain green and ethical mutual funds would probably increase and reach the general public and investors to a higher extent if broader groups of actors, not least SMOs and EMOs, were invited to play a larger part in deliberations about criteria. Such processes of convergence are so far only taking place modestly, through the large trade organizations of mutual fund companies where EMOs/SMOs are invited and consulted, although we have not yet seen examples of EMOs/SMOs – or other actors from outside the for-profit private sectors – having any comprehensive influence here. However, in an attempt to gain investors' trust in the SRI companies, a large number of such companies initiated the AICSRR, which supports the use of a standard of green and social screenings among SRI companies (see Chapter 5).

Environmental declarations are often framed in transparency terms and marketed with arguments that they include more detailed, substantive information about various aspects of the production processes. Calls for more comprehensive declarations in preference to labels are often based on the claim that declarations are better in line with the development whereby expert knowledge is growing and spreading among broad groups of the population in modern societies, both end consumers and professional buyers. In contrast to labelling, the alternative of more extensive information is framed as neutral and objective; the declaration is framed as a choice in favour of increased transparency. It is up to the receiver of the information to evaluate the information. In the electricity sector in Sweden, the environmental declaration is an alternative to the eco-label. Yet, the replacement of eco-labels with detailed environmental declarations has been met by criticism that concerns the risk of actually obscuring risky practices and giving the receiver the impression that the information is neutral and objective. Elsewhere, we have therefore argued in favour of distinguishing between different 'layers of transparency' of information and debates. Higher layers of transparency go beyond the detailed declaration of substantive facts, and add information as well as dialogues about decision-making procedures, the frames within which the principles take place, and so forth (Klintman & Boström, 2008).

The goals of transparency and environmental benignity are often confusing to investors and consumers with green and ethical ambitions. In the green and ethical fund case, merely because a company chooses to be transparent about its environmental and social impact does not – of course – mean that it is environmentally or socially progressive in its production, although companies frequently present their environmental declarations in that way. In certain sustainability indexes, environmental and social reporting, it is the only criterion for being included. The New Economics Foundation accordingly notes, 'we reward a company which manufactures harmful chemicals by placing it at the top of the Dow Jones Sustainability Group Index. Why? In part, because it produces a social report' (2000, p. 3).

In relation to such criticism, most labelling advocators maintain that transparency is necessary but insufficient for establishing credibility. They demand institutionalized supervision, particularly when there are good reasons to mistrust certified companies (as there always are according to some stakeholders). A common precondition for the possibility of making supervision effective is that the party that conducts this supervision has to be economically, organizationally, and cognitively independent of the producer or seller. A labelling scheme therefore normally requires that a certification body conducts auditing and on-the-ground inspection regularly, in order to verify compliance with standards. A standard-setting ideal, *auditability,* is relevant to our discussion here, and it could be understood in a broad sense. Power (1997, 2000) distinguishes between *auditing* – 'control of control', that is, control of self-control arrangements – and *inspection* – direct control. Because control of licence holders in eco-labelling can include both first- and second-order control, the term auditability covers both here. If stakeholders mistrust each other, a programme with auditing, inspection, and certification may help as an external reference point creating opportunities for mutual reflective trust. However, the auditing and inspection procedures need to be trustworthy as well. The subsequent debate in the seafood case reveals challenges involved in this regard (see also Boström, 2006a).

How could credible certification and auditing be accomplished given the huge lack of confidence expressed by EMOs, journalists, policymakers, and other policy analysts, in the whole fish sector? A great deal of the fishing sector's poor reputation had to do with suspicions of previous cheating within the business. There is a commonly held suspicion that fishermen fiddle with legal requirements, a claim which is made in official reports (Hultkrantz et al., 1997; Nordic Council of Ministers, 1998) as well as in EMO journals (*Sveriges Natur*, 2001/06 'Fusket som tömmer haven'). It is claimed that fishermen report incorrectly about capture and by-capture, that they practise illicit fishing, and make use of loopholes in the regulation. If fishermen cheat even with legal requirements, why should we trust that they comply with voluntary rules? What would be required by the voluntary system in terms of auditing and verification?

There was hot debate during the standards development process about such auditing issues. The debate led to careful efforts within KRAV's project group where they tried to develop trustworthy standards criteria. In the standard proposals, the project group referred to grand principles, such as transparency and traceability, which they believed would improve auditability. However, they were also pressured to suggest concrete measures. Their main idea was in part to rely on various self-made documentation (about, e.g., by-catches and location of catch). A satellite-based auditing system, which authorities provide for the automatic reporting of the position and movement of fishing vessels, could complement such documentation. Information about whether and where fishermen are fishing would then be provided. Various stakeholders, however, strongly disagreed about whether or not this was sufficient. Some (e.g. Coop) found the auditing function to be the most important factor for consumer confidence in eco-labelling, whereas other parties (SSNC, MSC) saw it as the Achilles' heel of the entire framework. One objection was that documentation of by-catches is hardly effective if there is no supplementary 'on board' inspection. The project group discussed whether inspectors should accompany the fishermen on the fishing trip. However, the project developed a standpoint that it would be too expensive to conduct an on-the-ground (or -dock, -board) inspection. Still, even if that were affordable, it is far from self-evident that sample inspections would solve all problems

In a sense, the eco-labelling system exists because we mistrust the (conventional) fisheries in their environmental performances. Simultaneously, the fish that is sold with a green label to a certain extent must be trusted to really comply with the eco-labelling criteria. It is important to note that auditing is basically motivated by *a degree of mistrust* in those audited; complete trust would make auditing unnecessary. Yet, auditing requires a certain degree of trust in order to be workable, since auditing is at least partly dependent on information that only the actors audited have (Power, 1997). The risk of audit, according to Power, is not simply that it does not work, but rather the tendency of actors involved to trust, affirm, and legitimize the potential of auditing and control in an unreflective way, without really knowing whether, and when, it works.

The seafood labelling case revealed huge difficulties in the attempts to establish credible auditing. However, we should not conclude from this case that establishing auditability through effective and skilled supervision is impossible in this or all other labelling cases. It varies according to factors such as technology, existing regulatory inspection programmes, and the degree of existing trust in the business that is subject to labelling. There are additional general problems, however. Such problems have to do with the type of eco-standard involved, and how the certification practice is organized. When Mike Power (1997, 2000) studied management systems, he noted problems connected with auditing. The problems are, he holds, due to dynamics in the certification practice. There is a preference for approving rather than rejecting, because the certification body wants a good relationship with clients in its market (see also Kerwer, 2008). Furthermore, there may be arbitrary interpretation of certification criteria or information asymmetry (to the advantage of the certified party). Other problems, despite third-party verification of standards, can be that auditing is conducted within a short timeframe (Newell, 2005, p. 553) or shortage of epistemic or material resources (Meidinger, 1999). In order to overcome such problems, the certification body should, according to Power (1997), develop an autonomous, discursive base (epistemic independence) in relation to those certified. The certification body should, moreover, be economically independent of those certified; it must not be negatively affected by rejecting or withdrawing certificates.

It is quite essential that the certification body be organizationally independent of the certified company. As to the American forest case, one of the most important initial criticisms of the SFI programme was that it did not have to be audited independently by outside organizations, as members could either audit themselves (first party) or have the AF&PA do it (second party). To address these concerns, the AF&PA changed its policy in 1998 to allow third-party auditing. Since then, although it is still voluntary, most of AF&PA's large industrial members have undergone such auditing (Cashore et al., 2004), which shows how forceful the norm of third-party auditing is.

Furthermore, the clarity and stringency of the standards may facilitate credible auditing (that is to say, the lack thereof may obstruct credible auditing) (Gulbrandsen, 2004). Performance-based standards with measurable and prescriptive standard criteria – in contrast to mere standards of procedures as in environmental management systems – strongly enhance the possibilities of conducting substantive auditing (cf. Humphrey & Owen, 2000). It can therefore matter significantly how specific eco-labelling criteria are formulated, which is a topic of much internal debate among stakeholders in labelling processes. EMOs tend to favour quantitative expressions that simplify inspection and prevent companies from finding ways to circumvent compliance. Companies, for their part, often favour qualitative expressions that allow for flexibility when practising the standard. Compare, for example, the following requirements: 'More old and big trees should be preserved at clear-cutting', and 'ten old and big trees per hectare should be preserved at clear-cutting'. The latter formulation would be easier to inspect and harder to circumvent, but might also be inflexible to shifting natural circumstances. Some warn that the focus on quantification draws attention away from other important but less measurable aspects (Thedvall, 2006; Lindvert, 2008). Gregory (2003, p. 565) reminds us: 'What is measured is what can most easily be measured'. Furthermore, this author asks the eternal question 'Is it better to be roughly right than precisely wrong?'

The impression of strict auditing and inspection can result in a simple and excessive trust, which may lead to businesses legitimizing their actions according to these conditions. One possible antidote to excessive trust in the auditing and certification practice is the inclusion of external stakeholders such as SMOs in the monitoring stage as well. Activities such as monitoring and evaluation can be broadly participatory and allow stakeholders to check actual performance (Lovan et al., 2004, p. 248; Van Rooy, 2004, p. 142). Besides using traditional means, for instance staging media protests, citizens, EMOs, and other SMOs can address concerns directly to standard-setting bodies or certification bodies or remind corporations to follow the standards properly. EMOs, in particular, play an essential role in such practices. Swedish EMOs that were highly involved in establishing the Swedish FSC label (e.g., the SSNC and the Swedish Ornithological Society, SOF) monitor the actual performance of forest companies. Sometimes these EMOs criticize companies for not complying with the standard criteria. This kind of stakeholder activity may be difficult for ordinary consumers to engage in, however. Our cases reflect this difficulty, in that we do not see the development of labelling institutions that allow consumers to check the actual behaviour of certified companies.

## Conclusion: towards mutual, reflective trust (among those involved)?

Dealing with matters of (mis)trust and credibility is at the heart of green labelling activities. In this chapter we have maintained that a potential quality that an inclusive labelling organization can add is an ability to mediate and overcome particularistic interests. Repeated interaction over time in organized networks that comprise a wide range of actors can result in common expectations about proper behaviour. It can also lead to a degree of mutual learning and mutual reflective trust. This chapter has analysed this theme further in relation to certain problematic circumstances (contested views and uses of science, cognitive authorities, and audits). Our analysis adds to the conclusion that actors involved in labelling tend to get closer to each other through labelling processes, something that almost all interviewees mention. However, our cases also illustrate the many struggles and setbacks involved in achieving such rapprochement as well as drawbacks and increasing mutual mistrust, especially, but not exclusively, from our US cases. We have to reach the conclusion that the development of mutual reflective trust may – at best – concern only the actors and groups *involved*. We see few indications that a broad group of politically concerned consumers gain directly from this repeated interaction, cognitive authorities, and auditability.

# 11
# Green Labelling and Green Consumerism: Challenges and Horizons

Many producers, consumers, and policymakers understand the positive value of green labels, and believe that green labelling must be part of the action repertoire in the struggle for a more sustainable society. In the concluding chapter – and based on the findings of this book – we want to discuss the preconditions that green labels offer for the greening and democratization of society. It should be clear, however, that we do not aim to draw firm conclusions about the specific ecological consequences of certain policy instruments. For one thing, we are social scientists, and analyses of ecological dynamics are beyond our task and our level of expertise. Furthermore, it is extremely difficult to assess the effectiveness of a specific policy tool because of the complex causal relationships between standard criteria and industry practices, including multiple direct, indirect, unintended, and mutually counteracting effects (cf. Stokke et al., 2005). Yet we can approach this question by investigating how *social* structures, processes, and actions can hinder or facilitate reflection on, dialogue about, activism regarding, and political collaboration in environmental problems – which we are trying to do in this chapter. In parallel, we are allowing ourselves to engage normatively in the issue, and to address some critical policy implications. Our intriguing point of departure, introduced in the first chapter, is that labelling is about translating social and environmental complexities into a simple, categorical label: a 'this is the green(est) choice' label.

The reader should not be surprised that we argue that green labelling has limits. Moreover, we believe that the imperative 'Consume Less!' is a more crucial and urgent task than is the conscious choice among green products – at least for all rich and rapidly developing countries in the world. Labelling relies on the conscious choice strategy, and may even be contributing to difficulties for the consume-less strategy in entering green political agendas. We maintain, moreover, that many producers, consumers, and policymakers are overly optimistic about the potential for green labelling to solve various issues; others may be overly pessimistic, however.

The very fact that green labelling has certain limits is not a valid reason for rejecting it. In this chapter, we argue that green labelling has interesting potential, but not necessarily for the reasons that many optimists emphasize. Green labelling also has weaknesses, we argue, but not necessarily those that many critics address.

To us, labelling is part of the repertoire of policy instruments, eco-standards, and other consumer strategies. Labelling alone cannot solve any environmental problem. In this book, labelling practices have been analysed in their discursive, organizational, regulatory, political, and transnational contexts and processes, and any discussion of the potential of labelling has to account for such contexts and processes. We structure the discussion according to the four broad themes that we introduced in Chapter 1 (see Table 11.1).

*Table 11.1*  Four labelling themes

| Theme | Politics and science | Reflective trust | Symbolic differentiation | Relationship between production and consumption |
|---|---|---|---|---|
| | Does politics distort labelling? | Should labels be trusted? | Should integration or differentiation be used? | Are the labels in accordance with the concerns of consumers? |
| **Myths and Misconceptions** | 'Politics distorts labelling.' 'Labelling is objective and neutral; it reflects or should reflect pure science.' | 'Labelling relies on the (simple) trust of consumers.' | 'Green labelling cannot be effective because it cannot fully scale up.' 'Green labelling should scale up.' | 'We should prevent competition among labelling schemes because consumers are confused by so many schemes.' 'The simpler and more clear-cut the labels are, the better.' |

Continued

*Table 11.1* Continued

| Key issues and horizons | Politics empowers labelling | Reflective trust empowers labelling | Symbolic differentiation empowers labelling | How can consumers be empowered? |
|---|---|---|---|---|
| | **Political consumption** a) How to maximize the consumption of labels vs. how to maximize 'responsible' consumption, following a political or ethical principle | a) Creating organizational forms and frames that allow reflective trust to be developed among broader audiences | a) Labelling can be effective and powerful – not despite but because of its 'marginality'. | a) Meta-politics and organizational forms and framings allow for consumer empowerment. |
| | b) Political empowerment outside of labelling institutions: 'the private is political' | b) Facilitating frame reflection | b) The power of labelling is related to the dynamics involved in symbolic differentiation. | b) Consumer empowerment includes cognitive, material, and emotional/ aesthetic aspects. |
| | **Sub-politics** a) Opening up core corporate economic and technological decision making to stakeholder input | c) Creating a forum for education and discussion – learning the limits and opportunities of labels, eco-standards, and other consumer tools | c) Managing and reflecting upon this symbolic differentiation (balancing between marginalization and main-streaming) | c) Labels cannot be attached to every citizen, which is a fact to be taken seriously by labelling agents when they orient to consumers. |
| | b) Stimulating co-production of politics and knowledge, including science and other forms | | d) Developing labelling as supplementary to other policy instruments, eco-standards, | (d) Efforts to close the gap between pro-duction and consumption could include |

Continued

*Table 11.1*   Continued

| Key issues and horizons | Politics empowers labelling | Reflective trust empowers labelling | Symbolic differentiation empowers labelling | How can consumers be empowered? |
|---|---|---|---|---|
| | of expert and lay knowledge | | and consumer tools (and vice versa) | issues of consumer representation; preservation of SMOs' critical distance, the simultaneous marketing cons; the participation in meta-politics; and the framing of the labelling instruments as 'communicative tools' rather than 'information tools'. |
| | c) Hopes and disappointments with inclusiveness | | | |
| | | | e) Acknowledging competition among labels and other eco-standards | |
| | **Meta-politics** a) Unveiling the sub-politics of labelling | | | |
| | b) Debating labels vs. other eco-standards, consumer tools, and policy instruments | | f) Acknowledging that the symbolic differentiation is arbitrary and relative – a topic for meta-politics of labelling that can encourage development of reflective trust | |
| | c) Stimulating frame-reflection | | | |
| | d) Creating collaboration among public and private authorities | | | |

## Politics empowers labelling

In Chapter 1, we addressed the misconception that labelling should be rejected because it is impregnated with politics, and cannot be neutral and objective. The opposite but equally problematic position is that labelling is good because it is objective and neutral, in the sense of being free from any political agenda: labels simply show us the greenest choice. However, we have wanted to challenge this common view that green labels are or should be reflectors of a pure and neutral (scientific) knowledge. Both expert and lay knowledge are needed, as well as strategic and ideological visions. Indeed, it is political envisioning that ultimately empowers labelling.

As we maintained in the first chapter, green labelling is an inherently political activity, and the political nature of labelling appears on various levels – on the levels of political consumption, subpolitics, and metapolitics. Science, too, appears at all these levels, and one could argue from a normative standpoint that science, along with other forms of knowledge, should be part of each of these levels. Science, however, should not be seen as a self-evident authoritative centre, but should be treated critically and reflectively. The next sections examine the ways in which politics empowers labelling at all these levels.

### Political consumption

It is not self-evident that the purchase of, for example, organically labelled vegetables should be seen as a political act or as a responsible act in the sense of being guided by an ethical intention (Boulanger & Zaccai, 2007). We may argue, to the contrary, that there are mixed motives behind the purchase of organic vegetables. Some motives are more political, in that the consumer wants to eliminate environmentally objectionable practices, whereas other motives concern individual health and product quality. Indeed, the existence of such mixed motives largely explains the relative strengths of various types of green-labelled products, particularly in the food area. Green electricity is a more abstract concept, and it is not as clear how it connects to individual self-interests, except as an indicator of a green image and political identity (Lindén & Klintman, 2003).

Yet if there is a wish among many people to look for more flexible, spontaneous, everyday channels through which they can express political engagement and responsibility, and if traditional political parties are seen to be inert and to have difficulty integrating new problems into their ideologies and actions, green labels can be used in accordance with the identities and agendas of political consumers (Micheletti, 2003). Such politically minded acts can be important drivers in green labelling. On the surface, this appears not to be the case in certain sectors. In forest certification, for example, we have seen that 'real' political consumers play a marginal role; it is rare in this sector that everyday consumers explicitly demand eco-certified furniture. Still, it is the potential threat of negative EMO campaigning and consumer boycotting that largely explains the willingness among DIY and other retailers in the forest market to push for certification (Gulbrandsen, 2006). Thus, even if there is a good reason not to exaggerate the capacities and willingness of a broad group of consumers to engage in everyday green consumerism, about which many academics remind us, one should by no means disregard the potential consumer power. As long as businesses are concerned about their reputation and brand, the 'imagined' and 'represented' consumer can play decisive roles (van den Burg, 2006). Even if consumers stick with conventional shopping behaviour, they can nevertheless play latent roles by being mentally prepared to discriminate among products because of a concern for and solidarity with other people, animals, and the environment.

To be sure, there are, arguably, several good reasons for labelling advocators to frame the win-win scenarios of shopping green (organic is good for the environment, animals, taste, safety, quality, health, nutrition, and sense of well-being). If the goal is strong market impact and more businesses converting to green practices, it is wise for labellers and marketers to reflect upon possible mixed motives. On the other hand, if one adopts the goal that a maximum number of consumers should act 'responsibly' by deliberately following a political or ethical principle, this big, broad strategy is not necessarily the best one (cf. Boulanger & Zaccai, 2007). Consumer insight and an increasing potential for reflection (Chapter 4) should be at least part of the goal.

There already is an embryonic infrastructure for creating more political and ethical consumers in the full sense of the terms. Yet we think it is critically important that this infrastructure be built up not merely by labelling institutions and other eco-standard setters. Perhaps it is difficult for labelling administrators to be engaged in issues other than marketing, credibility issues, policy monitoring, and fine-tuning of standard criteria. There must be other groups without vested interests in a specific labelling programme that engage in debate over, and the dissemination of, information – specifically about labelling and more generally about green consumerism. Hence, the political empowerment of consumers is more likely to be developed elsewhere, prominently within civil society or within educational institutions. As Giddens has suggested (1991), this empowerment concerns identity politics or life politics. Actors such as teachers, the media, EMOs, consumer associations, and other social movement actors will have a potential key role in this regard.

One of the key issues in which new social movements have been engaged in recent decades is telling people that the private is public, that lifestyle patterns and consumer choice matter, that every single act of consumption matters in precisely the same way as one vote in the political system matters. Still, to claim that the private is political is far from commonplace in everyday thinking, because we have learned in Western cultures that we are and should be self-interested actors as soon as we enter the market. If green labelling is closely connected to green consumerist agendas, pushed by various social movements, and if related notions of politics and ethics are communicated in a deliberative manner, its substantive (green, ethical) and democratic potential is likely to be significant, we argue. This issue is discussed further in the next section.

## The subpolitics of labelling

Ulrich Beck (1992, 1994) understands subpolitics to mean the opening up of economic and technological decision-making in the closed corporate world to broader stakeholder influence. Labelling fits this description perfectly, because the setting of standard criteria involves struggle, negotiation, and communication among a broad group of organized actors. Labelling efforts indeed concern core corporate activities, including economic and technological issues (although free trade rules may impinge upon the possibility of discriminating among technologies; see Chapter 7). In labelling processes, external stakeholders have an opportunity to debate the sustainability of certain corporate practices. In this sense, green labelling goes much further than environmental management systems (EMS) and environmental reporting and declarations, as these eco-standards do little more than address core corporate administrative *procedures*. One could even say that EMS is built on an obsession with procedures, or on management doctrines that are already commonplace and uncontroversial in the corporate world. Labelling, more than any other form of eco-standard setting, can potentially trigger subpolitics in Beck's sense of the word.

We have noted that there are always combinations of politics, science, and other forms of knowledge in labelling practices. Although science and other cognitive authorities appear to be grand standard-setting ideals (cf. Chapter 10), all our cases indicate that actors involved in the labelling practice are well aware of the political nature of labelling. They participate in communication and compromises, developing a reflective understanding of the process, which includes the indispensable strategic and political input. This reflectivity also nourishes an understanding of the uncertainties involved: that science is necessary but insufficient, and that the translation of social and environmental complexities into a simple categorical label necessarily includes pragmatic compromises and a degree of arbitrariness. Yet judgemental relativism sometimes mixes with epistemic absolutism, which is arguably inconsistent.

Contrary to the common contention about the importance of simplifying labelling and standards, a key notion in this book is that a more reflective view of science and a broad array of knowledge can enhance the potential of green labelling to reduce various environmental problems (the 'substantive' potential). It is difficult for the members of a single organization or a single discipline to know the ultimate combination of strategies to use when trying to deal with environmental problems from every angle. On this basis, several scholars who examine environmental policymaking more broadly have argued for more inclusive policymaking (see, e.g., Lafferty & Meadowcroft, 1996; Mol et al., 2000; Tatenhove et al., 2000; Pellizzoni, 2004). Through constructive dialogue, reflection, negotiation, and compromise, groups with different concerns, knowledge, and experiences may be able to shed light on different aspects of the problem and stimulate reflection, while taking responsible measures. Labelling is interesting in this respect, because several labelling programmes involve a great array of groups in the policymaking. Yet hopes for inclusiveness have their vivid opposite in fears and disappointments, founded on cumbersome decision making, stalemates, overrepresentation and underrepresentation of certain actor categories, power shifts, and painful compromises (Chapters 9 and 10; see also Boström, 2006b). Furthermore, despite the inclusiveness, there is one group in particular that is absent: the (political) consumers.

What may be surprising is that labelling initiators often fail in two respects. They fail to develop novel ways of representing political and ethical consumer concerns, as such concerns are seen as being represented by other organisations or retailers. Furthermore, the initiators fail to disseminate their reflective understanding of the co-production of politics and knowledge to a wider audience. The labelling is marked by subpolitics – by the opening of core corporate economic aspects to stake-holder dialogue and negotiation. However, the subpolitical nature of it remains largely hidden. The subpolitics of labelling is concealed behind the label, behind an expert-oriented rhetoric, behind talks of objectivity, neutral information, and scientific evidence. The subpolitics rarely involves wider audiences, and does not involve concerned consumers. If subpolitics concerns the opening of core corporate economic and tech-nological decision-making to a wider group of stakeholders, there is still a need for opening *the subpolitics* to a broader audience of concerned consumers. This requires the metapolitics of labelling.

## Metapolitics of labelling

Labelling is also empowered or obstructed by politics and policymaking that take place 'above' (or before, or in parallel with) labelling processes: the metapolitics of labelling. Whereas subpolitics concerns debate and negotiation over labelling principles, procedures, and criteria, metapol-itics concerns debate and discussion about labelling and its relationship to eco-standard setting and to green consumerism in general, and to reflections on the frames underlying various labelling programmes. Metapolitics would involve reflections on the general conditions of labelling. (A book such as this one would hopefully stimulate such reflections.) The metapolitics of labelling could potentially play an important role in both the empowerment of individual political con-sumers and the unmasking of the subpolitical nature of labelling proc-esses. Such politics could involve many types of actors and initiatives. Just as pupils learn about the political system at an early age, so can they be taught the principles and practices of political consumerism. Given their key roles in this process, social movement actors would have a natural role in arranging various types of adult education (evening classes, study circles) and practical applications on the subject. Radio and television could broadcast programmes with debates on con-sumer power in which various consumer tools and strategies were com-pared. State actors can assume a strong role in the metapolitics of voluntary regulation such as labelling.

We have seen, particularly in Chapter 7, that political (policy) context creates favourable or unfavourable structural or cultural conditions for labelling. On the one hand, the state may be too involved, as in the case of organic standardization, in which actors within the federal authori-ties in both United States and the EU seek to take control of it. This centralization of the eco-standard setting appears to be related more to strategies to facilitate trade and the internal market than to promoting green political consumerism. Fear of the pluralism of labels is at play here. On the other hand, we have seen that state actors can provide practical and symbolic support for green labelling initiatives in a more creative fashion. Compared to the organic case, the Swedish forest case illustrates a more collaborative atmosphere between state actors and state regulation, on the one hand, and voluntary forest certification initiatives on the other. It is clear that EMOs, among other actors, demand more command-and-control regulation in this sector. Yet state rules have facilitated and provided a platform for the development of private rules. State actors demand that the industry 'voluntarily' take relevant measures to protect the environment; otherwise, they may be required to act. Consequently, there has been a need within the industry to maintain good relations with the state. Certification has been a method for showing that they really shoulder their responsibilities. Although Charles Lindblom characterized the state as having a 'strong thumb but no fingers', the type of jointly adapted regulation we are discussing here clearly indicates that there is a role for a strong thumb as well as some fingers (Lindblom, 1977).

The state has another opportunity in the metapolitics of labelling: state officials could learn from new knowledge, ideas, and experiences gained from labelling initiatives and could make use of such expertise in their own policymaking and rule-setting. It appears, however, that this opportunity represents potential more than it represents actual practice. From our cases we see, particularly in Sweden, varying degrees of willingness among state actors to learn from private initiatives.

Metapolitics would probably look different in the United States than it does in Northern Europe. In the adversarial political culture of the United States, where we find many more antagonistic debates, oppositional anti-consumerist movements, and tensions between public and private policymakers in various labelling processes (Chapters 2 and 7), the most challenging strategy would be to find constructive relationships between private and public actors. Such metapolitics could visualize the deep mutual mistrust among groups and the epistemological and ideological crossovers that are commonplace in this policy context. Metapolitics would consider how regulatory frameworks, rather than restricting the space for green consumerism, could facilitate progressive private labelling or other beyond-minimum-law-rule-making initiatives. While there is scarcity of constructive relations here, protesters should not cease mobilizing people and targeting public policy, because it is evidently the case that protests matter. We have seen, particularly in organic labelling, that the normally restrictive American attitude towards regulation actually mixes with a certain political culture of 'readiness to regulate', as a response to protest campaigns and public input.

In the Swedish consensual political culture, we have found a good climate in several sectors for the development of a readiness to negotiate, for development of mutual reflective trust among a broad group of stakeholders, and consequently for effectively carrying out inclusively organized labelling processes. However, a more challenging task for the metapolitics in this country would be finding ways to facilitate frame reflection. Whereas a consensual political culture can enrich a debate by allowing nuanced argumentation within labelling networks, an adversarial political culture allows the introduction of multiple frames. A key task for metapolitics in Sweden would be to bring in novel frames in various public forums where various actors may visualize and discuss their fundamental epistemological and ideological assumptions.

Knowledge about pathways towards labelling in different policy contexts – and the pros and cons of each – would enrich such metapolitics. Much of politics beyond labelling practices also takes place within and among organizations such as ISEAL and IFOAM. We believe that such organizations will have to play key roles in the continuous transnational policy monitoring for carving out a regulatory space. In part, these efforts will include struggles against other general doctrines (embodied, e.g., by the WTO and ISO).

## Reflective trust empowers labelling

Our next strategy was to challenge the common view within the green labelling literature and policymaking that green political consumerism necessarily is, and should be, based on a *simple* trust among consumers. On the contrary, we claim that reliance on simple trust threatens to undermine green labelling in the long run.

To be sure, labels are substitutes for our senses. Thus, green (political) consumers must place some type of trust in the label, including the general labelling process and its organization; but what kind of trust is implied when this view of the relationship is addressed? We notice that labellers and stakeholders involved in labelling processes seldom consider this matter; rather they are typically geared towards stimulating a simple trust, something which is closely connected to epistemic absolutism.

In the organic case, the transition from 'confusing' and multiple schemes in different states to one national standard is completely consistent with the ideal of simple consumer trust. Moreover, it is common rhetoric in the organic movement that the 'stricter' the labelling criteria, the more trustworthy the label, even though it may be technically difficult to check that the stricter criteria are followed. There is little surprise, moreover, that all the cases used in this book and the comparisons across countries reveal the strong significance of messages that call for simple trust in green marketing and labelling. Brief marketing information is directed, almost by definition, towards the simple trust of potential consumers and investors. Browsing the websites of SRI funds from Europe and the United States, for instance, one finds slogans such as 'Do you want to invest in the weapons industry?' (www. banco.se), and 'Invest for your future while helping to build a world of peace and justice' (www.domini.com).

We have seen in Chapter 10 that several labelling arrangements help to encourage the development of mutual reflective trust among the participants involved. At other times, labelling debates entail an increasing mutual mistrust. Both policy contexts – a political culture of readiness to negotiate, for example – and skilful organizing combined with repeated interaction may facilitate the development of a mutual reflective trust. External reference points, such as cognitive authorities, may fuel a process of trust as well. Although it can be difficult to change the context in which labelling agents operate, they may elaborate skilfully on the above-mentioned policy contexts and reference points. Swedish seafood labelling, for example, has demonstrated that labelling agents designed procedures with the intention (i.e., partly reflecting previous power asymmetries and mutual mistrust) that these procedures would eventually lead to a friendlier general atmosphere. If successful, the gradual development of mutual reflective trust establishes a social capital that can be exploited in subsequent policy processes.

It is important to emphasize that the mutual trust of stakeholders is far from being a blind, unreserved, passive, and simple trust. Mutual reflective trust means that participating actors are aware of the possible unspoken agendas of other parties. We should never expect complete mutual trust to be developed among such different types of organizations as EMOs and corporations. Instead, mutual reflective trust implies a degree of suspicion, scepticism, and mutual checking, while actors simultaneously learn that it is possible to engage in a dialogue and develop mutual expectations.

As mentioned, however, only modest efforts have been made by labelling actors, other stakeholders, and policymakers to stimulate such trust relationships among the broader public. Mutual reflective trust is restricted to participants directly involved in labelling. *Reflective trust* is a trust that the standards are capable of improvement, and that consumers and a wide group of stakeholders are needed in these processes of continuous modification – as individuals and as members of organizations. If that were the case, we assume that consumers would be better equipped to comprehend the inherent knowledge fallibility, the ideological diversities, and the political priorities of labelling policies. Such reflective trust would be facilitated by the opening up of the subpolitics of green labelling by way of metapolitics.

Encouraging a combination of commitment and sound scepticism among concerned consumers would have to entail notions of how better to involve or represent various groups of consumers, and ways to facilitate frame reflection. It would involve reflections on how to stimulate consumer insight and consumer influence, in addition to the trust dimension (see Chapter 4).

In our view, the development of carbon labelling (sometimes called $CO_2$ labelling), in all its complexity, concerning how to measure 'carbon footprints' from various goods and services (see, e.g., *The Economist*, 2007, 383(8529), 90) is much in need of consumer groups learning about such schemes and becoming actively engaged in influencing them. Without inviting a wide range of consumer organizations that have good contact with various consumer groups, the complexity and value bases that are inherent in such schemes are unlikely to stimulate reflective consumer trust. Even less likely would be a major change among consumers and households in their daily practices based on such unengaging schemes. Again, forums are needed for education, discussion, and debates in which consumers can learn about the limits and opportunities of labels, eco-standards, and other consumer tools.

To be sure, this all has similarities with Bowker & Leigh's recipe in their ground-breaking work 'Sorting Things out', where they analyse classifications in general:

[a] key for the future is to produce flexible classifications whose users are aware of their political and organizational dimensions and which explicitly retain traces of their construction. (Bowker & Leigh, 1999: p 326)

They also – very plausibly – end their book by contending that 'The only good classification is a living classification' (Bowker & Leigh, 1999, p. 326). Still, we find it necessary to add that the only living classification is the one which also helps to move users, or at least user organizations, beyond mere awareness, into a more active engagement (see below in 'How to empower the consumers').

## Symbolic differentiation empowers labelling

As we have maintained throughout the book, green labelling relies – by definition – on symbolic differentiation. Based on this claim, we challenge a common critical view that green labels cannot be effective because of their limits to 'scaling up'. Some researchers and policymakers say that the green labelling strategy is an ineffective means of tackling environmental problems, because green-labelled products will appear only as small niches. Should the goal not be the reform of entire industries towards sustainable practices?

In contrast, we argue that green labels can be effective because green-labelled products appear as a 'top' niche within markets. To be sure, the goal is not that the labelling instrument alone should reform entire industries. It could, however, help to initiate a process in that direction. Differentiation is essential for the dynamics of green labelling. A great challenge for labellers is, arguably, mechanisms that counteract symbolic differentiation and that work in a mainstreaming direction.

The niche cannot be too small, however, or too detached from mainstream markets. It must be visual. It must be a niche that appears intriguing or, to a certain extent, threatening to a much broader audience, including important parts of the 'conventional industry'. Labels visualize and communicate 'the best choices' to consumers, but also in relation to many other audiences, including competing producers and a broad network of policymakers. Labelling must threaten to increase market share without absorbing entire markets. It is probably most powerful when it appears to be increasing. The observations of Jordan and colleagues are relevant to our discussion:

> Once a critical mass of businesses has applied successfully for an eco-label within a certain market segment, the remaining companies find themselves under considerable market pressure to seek the label for their competing product(s). (Jordan et al., 2004, p. 176)

### Why does the labelling strategy not have to 'scale up'?

For environmental or ethical reasons, it is sometimes desirable that entire product categories be withdrawn from markets. The labelling tool appears ineffective if the goal is to ban a whole product category; traditional command-and-control regulation is the typical means to such an end. Moreover, in its current designs and variations, labelling cannot tackle the urgent imperative 'Consume Less'. The possibilities of labelling are more modest, yet more substantial than some pessimists would have it. Admittedly, it is clear that we should be somewhat cautious in our forecasts. We do not want to echo the grand promises and hopes about green consumerism that are sometimes disseminated by various policy actors. Yet we assume that green values, norms, and ideas channelled through green labelling can indeed matter and can make a difference, for both green adaptation and broadened democratic engagement.

If we distinguish between the direct impact that labelling has on certified practices and other more subtle indirect impacts that the introduction of labelling can have, it is clear that symbolic differentiation is likely to play intriguing roles. Certification and labelling can, of course, affect the practices that they address directly: the certified practices. For the long-term credibility of a programme, it is probably essential that labelling organizations can report on or convince people of positive substantive outcomes of certification and labelling. Yet it is usual for such causal links to be extremely difficult to support with hard evidence that meets everyone's approval (cf. Stokke et al., 2005). Labelling agents tend, rather, to provide anecdotal evidence. Social and ecological complexities are normally too overwhelming. It is probably the case that labelling agents must rely on reflection, counterfactual reasoning, and persuasion by the use of concrete 'good examples' placed in their broader context, and by highlighting key factors in their reporting on the effectiveness of certification.

Dynamic indirect effects complicate the picture even more, but such effects make it more interesting. It would be drastically misleading not to reflect on scenarios in which indirect effects play a role. The introduction of labelling appears to be central, in that it has consequences far beyond the operation of single, certified businesses in the market arena. In all the sectors that we have studied, the introduction of labelling appears to have provoked or stimulated the introduction of new ideas, dialogues, and reflections on how to make any practice more environmentally or socially sustainable. Labels refer to new visions of practices by being based on systematic and coordinated experience and knowledge about such practices. It is apparent, for example, that ideas about organic agriculture have stimulated a great deal of green thinking in the conventional part of the industry and among public authorities – not least in Sweden, where organic agriculture is generally appreciated.

In this way, we should not evaluate results merely by counting market shares, converting rates, and certified hectares. We should try also to assess the many new – perhaps competing – initiatives that have been taken partly because of the first labelling initiative. Counterfactual thinking is important here. Would we have seen the SFI initiative in the United States without the FSC? It has often been said that the FSC is marginalized in the United States. However, such a statement can be made only if one ignores important dynamics in the labelling strategy. In fact, the FSC was seen by key industries in the United States to be threatening, and this is a fundamental reason why the SFI was established. Cubbage and Newman make a relevant comment about the competition between FSC and SFI when they hold that this competition has led to businesses adopting standards within both systems that are stricter than 'those that could have been achieved by government mandates' (2006, p. 271). Consequently, even the lower of these two standards goes well beyond legal compliance. The symbolic differentiation at play is part of the subpolitics that sets off new ways of thinking about green practices. This is not to say that there is always a 'race to the top'. Rather, the symbolic differentiation plays an important role in the development of new green ideas and framings.

Another version of green competition can be found in the US Green-e programme for electricity. There, the same labelling programme – Green-e – not only guides residential and commercial consumers to Green-e-certified energy companies; it also indicates the percentage of various renewable electricity sources that the various energy companies use and in what US-specific state the energy has been generated. It is fair to say that this clearly adds to symbolic differentiation, competition, and transparency, through a labelling programme certificate, without becoming more technical than most green consumers would be comfortable with.[96]

Hence, the label can also be seen and used as a template – a good example – that is related to a much broader field of politics and policy-making which, in turn, helps to create pathways towards sustainability. That labelling organizations compete over market shares and try to convince audiences about the competitors' flaws is not necessarily a bad thing, although many commentators think that competing labelling schemes create consumer confusion. Rather, such competition may stimulate broader public debate.[97] Labelling agents and stakeholders can use their voices in this (meta)politics, thus contributing the knowledge, experience, and interest that have been developed through their involvement in the labelling programme. Given our assumption that

symbolic differentiation is a key power mechanism in the labelling strategy, it becomes relevant to explore the question of compromises in the labelling strategy.

## What are the general compromises in the labelling strategy?

It follows logically from our insistence on differentiation that this strategy can never be a self-sufficient means for solving a specific environmental problem. Green labelling should never be seen as a sufficient strategy or instrument for dealing with environmental problems, but merely as one tool in the action repertoire. Indeed, this strategy is dependent on its opposite: the relatively unsustainable, poor, dirty, conventional, risky, and grey product. Other regulatory tools must deal with these poorer products, and, as soon as they become 'greener', labelling principles and criteria need to be revised if the entire programme is not to lose its 'political' strengths. Potentially, state actors could play a creative role in making use of the new ideas introduced by the labelling programme. From this platform of knowledge and experience, they may design tougher regulations addressed towards the non-certified share of the market. The state could, therefore, set a new, higher bar for what is acceptable in the market, whereas labelling agents would then be able to raise their standards in an upward spiral. Unfortunately, we are unable to report examples of such creative and dynamic interplay between state and voluntary regulation, so this appears to be, as yet, more of a potential than a reality.

Another fundamental compromise of the labelling strategy has to do with its reliance on market dynamics. Even if we wanted to stress the critical role of symbolic differentiation rather than integration and mainstreaming, the labelling strategy is contingent upon the relationship with existing market and industry structures – including the chains of production and distribution – in order to enable some visual market impact. Such integration normally requires one to make compromises in the environmental message. Products must be marketable. They have to look normal and cannot be too expensive. Environmental issues are often framed in holistic terms, organic agriculture being a good example. The dynamics of the marketplace force the labelling practice to be economically efficient, and this pressure may combat some of the ideals and values that once motivated its very establishment (Allen & Kovach, 2000; Barham, 2002). As Raynolds (2000) maintains, there is always a risk that the space for alternative trade, which was opened up by social movements, will be subverted by profit-seeking corporations that appropriate the values added by the labels without adhering to the movements' underlying social and environmental values. There appears to be a general trend towards mainstreaming in those labelling programmes we have analysed and in 'ethical markets' in general (Crane, 2005; see also Le Velly, 2007 on fair trade).

Such mainstreaming tendencies, we maintain, are more threatening to the power of labelling than are the difficulties inherent in 'scaling up'. For some product categories such as detergents, the Nordic Swan label appears on almost every product. However, there is a negative side to such success stories, for they create the appearance that the label becomes a symbol of the 'average'. Thus, the label becomes more of a licence for market entry than a trigger for green and ethical visions of alternative practices and technologies. Licences for market entry could probably be achieved more efficiently through command-and-control regulation, however, or through industrial self-regulation.

Another problematic side of symbolic differentiation was discussed earlier. Symbolic differentiation may, in practice, support big businesses and discourage small ones. Symbolic differentiation is, in one sense, arbitrary and relative. Furthermore, although actors may strive to find valid ecological criteria to discriminate between practices and technologies, the distinction between green and grey is always social and political in nature. Such arbitrariness may lead to an unintended and unfortunate distinction between big and small. In several labelling programmes, we see that it has been a challenge to involve small, often community- or family-based, businesses in the labelling process for a number of reasons (e.g., economies of scale, risk-averse attitudes, lack of expertise in management and marketing, and poor links to market structures). An important challenge would be how to transform the symbolic differentiation into something positive, in which small actors have the opportunity to appear among the 'good' guys. Again, the current patterns can be partially explained by over-reliance on and integration with existing market and industrial structures.

Symbolic differentiation is counteracted by various actors. Some greens would argue that all products should be green, and that there is no need to differentiate among products in this way: 'We should get rid of all unsustainable products.' Yet we would still insist that the labelling strategy requires that there be many 'conventional' products. If all products were 'green' we would not need labelling, and continuous reduction of environmental harm would have to be triggered by other regulatory tools. EMOs usually favour symbolic differentiation in practice, however, because they want to single out the 'top' business actors – thus favouring an exclusive label, using various boundary framings. Other players, such as trade associations, often want to label the entire industry as green – thus favouring a highly inclusive label while aggressively counteracting any attempts at symbolic differentiation. Similarly, when there are competing models, such as FSC versus SFI, actors favouring the EMO-initiated alternative have an interest in explaining why this model is superior to the others. Actors promoting the competing industry-led model may want to tone down the differences among labels, telling consumers that their label is also credible and that it reflects sustainable corporate practices (see Cashore et al., 2004, pp. 238–239).

Cashore and colleagues, however, push the argument that the relationship is reversed in relation to producers at the beginning of the production chain. EMOs want to convince producers that it is not overly difficult to comply with the 'best' standard, compared with a competing industry-driven standard, whereas promoters of the competing industry standard try to convince others about the costs involved when using the EMO-initiated standard (Cashore et al., 2004).

Hence, there is a political and rhetorical dimension in the communication about the differences. EMOs and other actors (e.g., business actors that approve of labelling for reasons of green image, market entry, and green premiums) stress the symbolic differences, but cannot exaggerate this point in their communication with potential producers. Marketing and frame-bridging efforts towards new producers prevent such excessive symbolic differentiation. Promoters of industry-driven competing models, on the contrary, stress the symbolic differences in relation to producers who may be willing to comply with the EMO-supported label, but tone down the differences in their communication with consumers.

Retailers stand in a mediating position. They do not have to bear the cost associated with environmentally friendly production (except for environmentally friendly distribution, but this has been a relatively marginal issue in labelling – at least before the recent renewed public concerns about global warming). What we see in our cases, therefore, is that they accept symbolic differentiation, which is especially true when green labels are an integral part of their efforts to develop a green public image and offer products to a segment of ethically sensitive consumers. Yet they also want considerable market share, sales volume, and a continuous supply over the year; furthermore, they want to frame labelled products as 'normal'. This, in turn, requires the prevention of excessive symbolic differentiation.

In sum, symbolic differentiation is an essential part of the labelling strategy – part of the subpolitics of labelling. We think consumer power and the potential power of labelling can be undermined by the same factors that tend to undermine symbolic differentiation. This is not to say that symbolic differentiation is always beneficial. As mentioned, there is a degree of arbitrariness involved in the differentiation, as a result of social and political dynamics. Again, this arbitrariness should be an essential part of the metapolitics, and frame reflective debates of labelling. Reflective trust can be achieved only if consumers learn about such dynamics. On the one hand, symbolic differentiation must be part of the management and strategies of labelling agents. They must build a capacity for the constant adjustment of ecological criteria in labelling schemes in relation to what other labellers, eco-standard-setters, and policymakers do. On the other hand, such strategic decision-making should also be transparent to those not involved in this game.

A final note is related to credibility (cf. Boström, 2006a). As repeatedly stated in this book, maintaining symbolic differentiation in labelling is an enormously challenging effort in terms of trustworthiness. Claiming that something is better for the environment than something else is fundamentally provocative and sets off an entire fabric of activities, which we hope we have illustrated, as in the following straightforward quote:

> In order to be credible 'ethical' differentiation requires a whole company effort ... Compare, for example, the difference involved in backing up a claim that a company is more 'socially responsible' than a competitor or a product 'more ethical', with the rather more straightforward task of claiming that a sofa is more comfortable, a drink better tasting, or a manufacturer better at designing stylish automotives. (Crane, 2005, p. 227)

## How to empower the consumers

How should the gap be closed – or at least narrowed – between the production and consumption of labels? All our research behind this book has led to our final claim: that the production of tools for green political consumerism does not adequately correspond to the hopes, thoughts, uncertainties, ambivalence, and reflective capacity of the 'typical' concerned consumer. We have maintained that the insistence on objectivity, simplicity, and neutrality generally seen among advocates of green labels does not support the development of reflective trust. There is a mismatch between the production side and the consumption side of green labels. Social and environmental complexities are typically translated into a simple and categorical label. In one sense it has to be so. Labels are simple and categorical. In principle, however, there is nothing that prevents consumers from being invited to deliberate about challenges and choices concerning the translation of complexity into simplicity.

Likewise, we would argue, the common fear among debaters of green political consumerism that the existence of many, partly competing, labels is confusing to consumers is largely based on a simplistic view of consumers. The fear of competing labels disregards the insight and influence of green political consumers. In the area of traditional politics, we seldom observe a corresponding fear that there are too many political views or interest organizations. The existence of mutually competing political organizations is generally seen as beneficial for democracy, and it is usually assumed that they correspond to the differing visions and hopes among citizens. Similarly, environmental problems are complex and uncertain – even more so when they are placed in social contexts. The solution must necessarily account for a breadth of vision and priorities. In the long run, the reflectivity and ambivalence of consumers could very well undermine the potential of labelling if its proponents stick with a simplistic notion of the consumer. We believe that it is critically important to bear in mind that not every consumer is concerned, not all consumers will identify themselves as political or ethical consumers, and not all consumers will ever be interested in expressing their visions in the supermarkets.

Perhaps the labelling instrument cannot and should not be designed to suit a consumer category that is inclined to make use of free-riding, that is cynical, or that includes a profound sense of powerlessness. Labelling agents cannot engage the entire population (but maybe indeed half of the population, as in Sweden). And it is among those who are engaged that we find people with a broader interest in politics and ethical thinking, as well as the most well-educated citizens. At the same time, these citizens are the most ambivalent ones, and the ones most reluctant to green advertising (Chapter 4). (For a critical discussion of this middle-class bias of political consumerism, see Klintman & Boström, 2006; Klintman et al., 2008.) Again, it is wise to see labelling as a tool that must supplement a wide range of other mandatory and voluntary tools, and as one form of democratic participation.

Attempts at closing the gap between the production and consumption sides within labelling in particular, and green consumer politics in

general, could be initiated from both sides. It has to do both with
enriching the labelling process and with empowering consumers.
Regarding the production of labels, relevant measures would be those
that concern metapolitics and the organizing and framing of labelling
processes. The critical issue is not necessarily to achieve the maximum
inclusion of consumer input in the standard setting. Rather, reflecting
upon the organizing and framing or labelling, with the purpose of closing the gap between production and consumption of labels, would
include the following issues:

- Critical evaluations of how consumers and consumer groups have
  been represented in the standard-setting procedures. (What, e.g., are
  the biases in seeing consumers represented by organized interest
  groups such as retailers and EMOs?)
- Continuous evaluations of and searches for novel ways of incorporating
  consumer concerns, including ideas about the ambivalences
  confronting consumers (surveys, focus groups, panels, Web-based
  interaction, etc.).
- Research aimed at developing ways of improving interaction with
  consumers. Such research would enrich the standard-setting, while
  helping consumers and professional buyers with interpretations and
  practical uses of standard criteria.
- Cooperation without co-optation to prevent power shifts that are
  advantageous only to specific players, such as retailers. This cooperation would require self-reflection among SMOs. It would require
  reflection on organizational design, involving such issues as how to
  involve various players, remembering that forms may shape (but not
  determine) standard contents. It would entail a critical eye on the
  development of such common frames as metaframes, which tend to
  create cognitive path dependencies in labelling programmes.
- The development of ways to market specific advantages and unique-
  ness with each label, simultaneous with information about the general reductions, limitations, and compromises that have been made
  during the standard-setting process. This development should include
  communication about labelling as 'only one step ahead'.
- Participation in the metapolitics of labelling, and (self-)critical assessments of various programmes and of the various framings underlying
  these programmes.
- Framing of the labels as 'communicative tools' rather than
  'information tools'.

The last point would imply that the goal of labels is to initiate a dialogue (in relation to competitors, policymakers, consumers, etc.), rather than to inform about 'best choices'. Seen in this way, consumers could have something to contribute. From this perspective, consumers are not seen merely as information-takers, who are expected to behave rationally once they receive information on buying options. Labels certainly are, and should be, used as information; but in addition, they could be seen as a way to open dialogue.

Engagement of consumers in such a dialogue would require their empowerment. If there is to be any hope that consumers will express their consumer power in the market arena, there must be ways to look beyond the individualistic view of consumers. Consumer empowerment could include dimensions that are cognitive (e.g., knowledge gained about the subpolitics of labelling), material (e.g., access to affordable and accessible products), and emotive/aesthetic aspects (e.g., attachment to products and services in which consumers see the meaning and aesthetic value of products based on the green and social implications of these products and processes).

Would closing this gap have ecological and democratic advantages? We cannot really provide evidence for the extent to which a closing of the gap would have positive ecological consequences. What should be counted as *positive ecological consequences* is itself subject to subjective valuation, and should be open to debate. The answer also depends on the goals that various actors have defined for labelling. Potential goals could include the attainment of maximum market impact, for instance, or a maximum number of responsible consumers. If the latter is the answer, closing the gap is virtually a necessity.

Closing the gap would have democratic advantages, as it would improve channels for everyone to express and talk about green political visions. As mentioned, however, green political consumerism must be understood as one among many channels. It will never engage all citizens, and citizens will never have equal cognitive, material, and emotional capacities for using this form of political participation. The material dimension is intriguing when viewed from a north–south angle. Its fundamentally private nature as well as its northern focus – it is also fair to say bias – has set certain limits on democratic participation (Raynolds et al., 2007).

Labels cannot be designed to reflect an absolute state such as 'good' or 'sustainable', although labellers sometimes frame the situation that way. This book has shown that labelling is always dependent on its inherently political nature; moreover, labelling is highly reliant on symbolic differentiation. This book has also argued that this fact, with all its nuances, should be open to consumers through active, dialogue-based communication, which should help consumers to develop their reflective trust and become better equipped to engage critically and reflectively with this policy tool. In the ideals of green political consumerism lies not only hope for more market shares; there lies hope for more thoroughly reflective consumers who observe, think, and shop with an open mind, while a darker green postconsumerist society, such as the one envisioned below, seems ages away.

> *Roll your cart back up the aisle*
> *Kiss the checkout girls goodbye*
> *Ride the ramp to the freeway*
> *Beneath the blood orange sky*
> *It's last call*
> *To do your shopping*
> *At the Last Mall*
> (Steely Dan, 'The Last Mall', on the CD *Everything Must Go*, 2003).

# Notes

## 1 Introduction: Green Consumerism, Green Labelling?

1. In Chapter 3, we elaborate on the definition of eco-standards and green labelling.
2. There are many such studies. See, for instance, Boström et al., 2005; Klintman & Boström, 2006. Harrison et al. (2005) discuss the notion of 'the ethical consumer', and a number of articles investigate behaviours, beliefs, and attitudes of consumers. Similar types of studies are found, for example, in Zaccai, 2007; Batte et al., 2007; and Teisl et al., 2002. It should be mentioned, however, that the above-mentioned volumes also include analyses of institutional and other circumstances behind green or ethical consumerism, which are important sources used in this book. Furthermore, Gallastegui (2002) uses a broader perspective by including the relevance of eco-labels from the marketers' perspective, whereas Grankvist et al. (2004) compare consumer preferences with regard to negative vs. positive labels. McEachern and Schroder (2004) study the potential for consumers to express their views about eco-labelling.
3. This use of the term epistemic relativism is borrowed from the school of critical realism (see, e.g., Bhaskar 1989, p. 23; Soper, 1995; Sayer, 2000). A basic idea of this school is that, whereas all beliefs are socially and historically conditioned and thus subject to change, there are often rational criteria for judging some explanations as being better and more useful than others. Nevertheless, it should be mentioned that our choice of certain concepts from critical realism does not mean that we refrain from taking a strong position in the endless realist–constructionist debate, nor does it imply that moderate versions of social constructionism could not offer equally useful conceptual tools.
4. We have studied US and Swedish organic food labelling, supplemented with other Northern European examples, to an equal degree. In the cases of forestry and electricity, we have had a certain imbalance between our own Swedish/European and our US data. For the US forestry case, we use only secondary data. In the fishery and paper cases, we use rich primary data, but they refer mainly to the Swedish context. As to our case of green mutual funds, the balance has been equal between our interviews across the continents. Yet, the main part of this case consists of a comprehensive review of green mutual funds internationally, and of the body of research on the subject. We have been conscious of all the asymmetries that have led us to make comprehensive collections of secondary data. Moreover, by being part of the international research communities on environmental sociology, consumer studies and standardization, we have received a great deal of help from our foreign colleagues with more thorough experience of our cases in

their respective countries. Previous publications from our case studies are mentioned in Chapter 2.

5. For an extensive analysis of the adversarial policy climate in the United States, with focus on the consumer role as political, see Cohen (2003).

6. The various documents have, of course, been essential sources of facts and figures about various aspects of the labelling schemes and their organizational arrangements. Moreover, the documents have enabled an analysis of public statements, positions, and arguments from various labelling stakeholders. The interviews in our study are crucial complements to the document analysis. We have conducted 120 interviews, lasting from 30 minutes to 2 hours and 30 minutes and comprising both specific and general questions. In addition to our interviews with consumers, we selected informants with extensive experience in dealing with issues in their respective fields, thereby enabling us to make use of their specific expertise. The informants were asked to give an adequate and balanced picture of the general views, attitudes, understandings, and conflicts about labelling issues in the organizations they represented. Accordingly, the informants we chose had to be well aware of their own professional network and organizational setting (often as directors, board members, and managers). We designed a specific interview guide for each interview (questions specific for the organization), although a set of general questions was addressed in most interviews. For example, we asked questions about access: whether they believed that certain actors – themselves or others – had been excluded from or included in the labelling practice and whether certain ideas and issues have been included or excluded. Key questioning themes concerned the interaction processes: for example, (a) whether the labelling project had been marked by conflict or a collaborative atmosphere; (b) whether it had led to common understandings and expectations or whether disagreements and controversies had continued or even increased; and (c) whether the distribution of roles among stakeholders in the labelling arrangement was regarded as fair and reasonable. There were also questions about strengths and weaknesses in the labels. Some interviewees had substantial knowledge of such contextual factors as existing regulations to which the labelling was related. In-depth interviews were conducted until the data indicated saturation, which is the established criterion for choosing the number of interviews.

7. The case of green mutual funds has been a bit more difficult to study 'backstage' than the other cases, due to the partial business secrecy surrounding the development of new green mutual funds. Therefore, we have supplemented a more limited number of interviews with observations of meetings involving NGOs, funds companies, and governmental agents in the United States and Sweden.

## 2   The Historical Context – Key Trends

8. Such guides still exist, for example, The Good Shopping Guide (Berry & McEachern, 2005; see http://www.thegoodshoppingguide.co.uk/, accessed 2008).

9. Ahrne & Brunsson (2004b) maintain that rules are useful tools for (1) influencing and governing actors; (2) facilitating interaction and coordination among actors; and (3) establishing and maintaining identity and status relative to others.

## 3   Green Labels and Other Eco-Standards: A Definition

10. We here echo Bowker & Star's (1999) claim that even seemingly neutral classifications – in our case technical details of green standardization criteria – create advantages for parts of nature, groups of animals or people, and disadvantages or suffering for others. This makes virtually all classification carry a moral weight. In turn, it becomes important for the researcher to analyse norms of classifications, and to suggest alternative normative principles.
11. The Blue Angel is a voluntary third-party scheme with the German Federal Environmental Agency and a multi-stakeholder forum, the Environmental Label Jury, in the governance arrangement. The jury makes room for a wide range of stakeholders, including consumer and environmental NGOs, industry, churches, and scientists. The Blue Angel sets out to label best environmental choice within markets for products and services (e.g., paper, computers, washing machines, public transport, car sharing).

## 4   The Consumers' Role: Trusting, Reflecting or Influencing?

12. In the case of fair-trade coffee, the magazine claims that 'the low price of commodities such as coffee is due to overproduction, and ought to be a signal to producers to switch to growing other crops. Paying a guaranteed Fairtrade premium – in effect, a subsidy – both prevents this signal from getting through and, by raising the average price paid for coffee, encourages more producers to enter the market. This then drives down the price of non-Fairtrade coffee even further, making non-Fairtrade farmers poorer.'
13. Firstly, we have shown elsewhere that most consumers are interested in more aspects of products than merely product quality and price (Klintman et al., 2008; Ekelund & Tjärnemo, 2004). Thus, there appear to be hidden groups of green political consumers that are not found if the questions are formulated too narrowly. Secondly, depending on how green consumers are defined – as those who consciously follow green 'principles' or have a low negative environmental impact – two distinctly different consumer groups are found (Klintman & Boström, 2006). Thirdly, it is very difficult, if not impossible, to separate green consumer choices that have been made out of self-interest or green interest, particularly in the food sector. Fourthly, what are seen as 'political' and 'green' consumer choices are largely based on differences in culture across countries. Eco-labelling and fair-trade labelling are currently highly relevant in Northern Europe, whereas labelling is less common in several other regions, for instance Southern Europe. In the latter countries,

local and domestic production may be intertwined with green and political consumption in a particularly strong way (see Kjaernes et al., 2007). Finally, research on green and political consumerism typically focuses on daily products, which are still largely the responsibility of the woman in the household. Thus, it remains to be seen in future studies whether a larger focus on painting, construction tools, and chemicals used in automobile maintenance can give a more nuanced view of male consumer patterns.

14. Research on such comprehensive consumption issues usually takes place in studies of lifestyles, household practices and in public opinion polls (e.g., Mont & Bleischwitz, 2007; Lindén & Carlsson-Kanyama, 2001). Combining theoretical frameworks of political consumerism with these other research perspectives could most likely generate fruitful research for the future.

# 5   Our Cases

15. Our case studies on organic food labelling are documented in various reports, articles, and book chapters; for instance Boström & Klintman, 2003; Klintman & Boström, 2004; Boström & Klintman, 2006a; Boström, 2006a; Klintman, 2002a, b; Klintman, 2006.
16. See http://www.demeter.net/, accessed 3 January 2008.
17. Source: KRAV's annual review for 2006. Electronically available at www. krav.se.
18. Our case studies on forest certification are documented in various reports, articles, and book chapters; for instance Boström, 2002, 2003b, and 2006a. For the US case we have used secondary literature, and particularly the work of Cashore et al. (2004) has been important.
19. See http://www.fsc.org/en/about/policy_standards/princ_criteria, accessed 21 December 2007.
20. Source: FSC News + Notes, Vol. 5(10), December 2007.
21. Calculated based on information retrieved from http://www.fsc.org/ keepout/en/content_areas/92/1/files/2007_11_23_FSC_Certified_Forests. pdf, accessed 21 December 2007.
22. The general FSC Principles and Criteria were further concretized in region-specific standards across the United States.
23. Our case studies on GM labelling debates are documented in various articles and book chapters; for instance Klintman, 2002a and 2002b.
24. Since this Directive, two regulations have been introduced: one on mandatory labelling and traceability (EC) 1830/2003 and one on GM food and feed (EC) 1829/2003.
25. Aside from mandatory GM labelling, however, there are clear signs that EU regulators and agbiotech opponents are partly moving in different directions. According to Levidow and Boschert, this is reflected in the strong variation of 'agricultural development frames' and 'coexistence frames' (how GM and non-GM agriculture could coexist) across EU regulators and agbiotech opponents (Levidow & Boschert, 2008:179).
26. Our case study on marine certification is documented in various reports, articles, and book chapters; for instance Boström, 2004b, 2006b.

27. The main part of our research on the electricity case is based on unpublished research conducted in 2004–2007 by Mikael Klintman with the assistance of Erika Jörgensen.
28. http://www.green-e.org/, accessed 12 October 2007.
29. http://www.epa.gov/greenpower/aboutus.htm, accessed 12 October 2007.
30. http://www.energystar.gov/index.cfm?c=about.ab_milestones, accessed 17 March 2008.
31. http://www.energystar.gov/, accessed 17 March 2008.
32. For instance, the RES Directive (2001/77/EC) aims to increase the share of electricity produced from renewable sources from 13.9 per cent in 1997 to 22.1 per cent in 2010 (EU, 2001). Eco-labelling schemes of various kinds, and at various administrative levels, are among several strategies to reach this goal. Yet, voluntary eco-labelling of electricity in the EU has a rather marginal position in the European portfolio of policy instruments. Other policy instruments that have gained a more central position include $CO_2$ emissions trading and green certificates (see Gan, 2007, p. 152). The EU Directive 2003/54/EC (EU, 2003) concerns common rules for the internal market in electricity. That directive states that suppliers of electricity must provide information about the energy sources for their electricity production and the environmental impacts (at least emissions of $CO_2$). This is the basic transparency requirement that electricity with a green label rests upon.
33. http://www.energylabels.org.uk/eulabel.html, accessed 11 October 2007.
34. For an overview of electricity with a green label in other European countries, see http://www.greenlabelspurchase.net/en-gps.html, accessed 12 October 2007.
35. See, for instance, Micheletti's case study on this labelling programme (2003, pp. 122 ff.).
36. Part of our research on the case of SRI funds is based on an unpublished research paper by Beatrice Bengtsson and Mikael Klintman in the spring of 2007.
37. In practice, these as well as other types of screens are usually combined and exercised in different steps. They rest on a qualitative rather than a quantitative type of analysis.
38. http://www.unpri.org/signatories/, accessed 17 October 2007.
39. http://www.ftse.com/Indices/FTSE4Good_Index_Series/index.jsp, accessed 20 November 2007.
40. http://www.sustainability-indexes.com/, accessed 20 November 2007.
41. The European level is not the main level for organizing or screening of SRI funds. Such practices as the launching of sustainability indexes, along with organizing and screening of SRI funds, take place to a large extent at the intercontinental level as well as within countries. Still, the European level is also significant. Although it is not an administrative part of the European Union, the European Social Investment Forum (Eurosif) is a pan-European non-profit group that consists of pension funds, financial service providers, academic institutes, research associations, and NGOs. The members are made up of Social Investment Forums (SIFs) in a number of European countries.
42. http://www.csrwire.com/sb/article.cgi/3197.html, accessed 20 November 2007.

43. This case study was conducted by research assistant Sofia Nilsson, supervised by Magnus Boström, and was reported in Nilsson (2005).
44. According to SIS Miljömärkning AB, the organization responsible for the Nordic Swan, 85 per cent of the public in the Nordic countries know what that label stands for (see http://www.svanen.nu/Broschyrer/VemLyssnarDuPa.pdf, accessed 28 September 2007).

# 6   Sceptical and Encouraging Arguments

45. See http://www.ota.com/about/accomplishments.html, accessed 4 January 2008.
46. See, for example, the Rural Advancement Foundation International's website: http://www.rafiusa.org/ (accessed 17 January 2003); and the Organic Consumers Association's website: http://organicconsumers.org (accessed 17 January 2003).

# 7   Policy Contexts and Labelling

47. On the most general level, we may expect green labelling initiatives to take place only in countries with established market systems. While we see markets in all societies, not all societies embrace or contain market systems. A market system is a 'system of society wide coordination of human activities not by central command but by mutual interactions in the form of transactions [between buyers and sellers]' (Lindblom, 2001, p. 4). To be sure, we can see, as an example, that fair-trade certification has been developed in several developing countries without established internal market systems. Yet, it is the market dynamics in developed countries that explains these fair-trade labelling initiatives. A supporting condition – albeit not a necessary one – is the existence of democratic political systems (representative democracy) in the country concerned. Although green labelling exists also in countries with very weak democratic traditions, reliance on global markets is fundamental, and basic democratic structures in the countries where the initiatives are taken appear to be an important factor explaining the initiatives. Not least of all, democratic traditions foster political activism, including environmental and consumer-related concerns that are basic engines behind labelling initiatives. Finally, on a general level, it is clear that we should expect green labelling initiatives to take place in relatively wealthy societies, with a culture of solidarity rather than egoistic individualism (Cohen, 2005). Being able to express political visions by consumer choice is, quite naturally, related to a level of income; but also to a relatively fair distribution of income. We hypothesize that in a society with more equal distribution of income it is more likely that people on lower income levels are willing to express solidarity in shopping behaviour; they are less inclined to demand action and responsibility only from the richest.
48. Why do we concentrate on these context elements? In part, we focus on such elements as appear central to our cases, elements which clearly concerned the people (informants) involved in labelling processes. Hence, we have given weight to inductive reasoning. The analysis of the context

elements must also be relevant to our key themes discussed in the book. And, in a more deductive vein, we were also guided by existing literature on policymaking, rule-setting, and governance in general. Previous literature has helped us to look at this set of factors. Finally, our focused comparisons have enabled us to identify special or general opportunities and obstacles.

49. The concept partly overlaps other relevant concepts such as policy style (e.g., Richardson et al., 1982; Liefferink et al., 2000) and political opportunity structure (e.g., McAdam et al., 1996). Our notion of context factors as a latent propensity to act in a certain way is in our view closest to the term political culture.

50. JEP refers to 'The type of policy arrangement ... both jointly formulated and/or implemented by the state and private actors and by having a voluntary element' (Mol et al., 2000, p. 2), which can include labelling.

51. Since the mid-1990s the government has gradually adopted stricter goals for the growth of organic production. Organic labelling is now part of the general political goal and strategy that 20 per cent of the Swedish arable land should be certified organic by 2010 (Swedish National Board of Agriculture, 2004). Hence, the government clearly signals that organic production and food labelling are part of a strategic political effort to make the whole of agriculture more 'sustainable' (e.g., prop. 1997/98:2; Swedish National Board of Agriculture, 2004). A similar, albeit less explicit and ambitious, position is apparent in the 2004 EU Commission 'European action plan for organic food and farming'. See http://europa.eu.int/comm/agriculture/qual/organic/plan/comm_en.pdf, accessed 1 February 2005.

52. It should be mentioned that quite progressive sustainability-oriented policymaking is taking place below the federal level in the United States, something which we exemplify in the green electricity case.

53. The 60 groups represent the environmental and scientific communities, and include advocates of the small farm movement and consumers' rights organizations.

54. For example, the global fish regulatory regime with UN conventions and agreements is often criticized for being unable to tackle efficiently the overuse of fish resources (Porter et al., 2000; Stokke, 2001). Yet, despite lack of binding regulation and tremendous difficulties dealing with problems, such as overcapacity in the fishing fleet, a common global understanding is emerging of the need for improved fisheries management and conservation of marine biodiversity. The 1995 FAO Code of Conduct for Responsible Fisheries and its technical guidelines gave international support to improved fisheries management. It emphasized the importance of achieving sustainability objectives through market-based measures (Deere, 1999). Likewise, FSC developed out of strong criticism of the failure among existing IGOs to counteract effectively such problems as global deforestation. For instance, the UNCED in Rio in 1992 failed to establish a binding forest convention. However, as the establishment of FSC took place in parallel with the UNCED, several ideas permeating this event were repeated in the FSC framework, for example the emphasis on combining environmental, social, and economic objectives and the insistence on establishing an organization that balances the interests of the South and the North (Elliot, 1999)

55. These principles are formalized in the WTO Agreement on Technical Barriers to Trade (TBT) and the Sanitary and Phytosanitary (SPS) Agreements.

56. The Nordic Swan suffers from related problems. As regards the paper case, informants from the Swan thought it was more difficult to convince Swedish producers to use their label, because the companies in the pulp and paper industry have become less Swedish due to acquisition of companies. The contact persons are not always Swedish citizens, and they are less inclined to see the unique benefits of a Nordic label (Nilsson, 2005).

57. Menz (2005) concludes that there are still important impediments to green electricity in the United States. These include price distortions for fossil-fuel-based electricity. According to Menz, environmental regulations or taxes that are much more stringent, and that take place at a federal level rather than locally or on a state basis, are needed in order to create more rapid development for green electricity markets.

58. Meta-organizations are defined as organizations that have organizations (not individuals) as members (Ahrne & Brunsson, 2008).

59. See http://www.gen.gr.jp/index.html (accessed 23 January 2007). GEN presents itself as a non-profit association of third-party environmental performance labelling organizations. Its mission is to assist its members (about two dozen member organizations, including the US Green Seal, the European Commission DG Environment [EU flower], Germany's Federal Environmental Agency [Blue Angel], the Nordic Ecolabelling Board [the Nordic Swan], SSNC [Good Environmental Choice]) and other eco-labelling programmes and stakeholders, to engage in dialogue and debate with other kinds of policy actors, and to improve the credibility of eco-labelling programmes worldwide. It also aims to foster cooperation, information exchange, and harmonization among its members.

60. See http://www.isealalliance.org/index.htm (accessed 23 January 2007). Full members are FLO, FSC, IFOAM, MAC, MSC, SAI, and Rainforest Alliance (see list of abbreviations).

61. See Seippel (2007) on the environmental movement in Norway and Boström (2001, 2004a, 2007) on the environmental movement in Sweden.

62. In the 1980s, upscale supermarkets specializing in organic food appeared. These include Whole Foods Market, Bread and Circus, and Wild Oats, all of which have stores in many states (Boström & Klintman, 2006). However, the previous market division between organic and non-organic food stores is becoming much less clear-cut. In 2003, health and natural food stores accounted for 47 per cent of the organic food sales. Conventional mass markets sold 44 per cent, with direct sales through farmers' markets and coops, food service, and exports making up the remaining 9 per cent (OTA, 2004). Moreover, in 2005, The Organic Trade Association in the United States conducted a survey, called 20 Year Organic Survey Questions. To the question 'Where will organic products be sold in 2025?' a majority of respondents representing various organizations assumed that organic products ' will be sold anywhere and everywhere' (OTA, 2005).

63. The terms 'sociomateria' and 'sociomateriality' refer to the interwoven nature of materiality with society, a relation that, for instance, Orlikowski

(2007) and Law & Urry (2004) argue needs much more attention in future research.

64. In terms of policymaking the Swedish governmental consumer agency (Konsumentverket) in its evaluations has drawn the conclusion that the lack of separability between 'green' and 'grey' electrons is highly problematic from a legal point of view (cf. Lindén & Klintman, 2003), although such legal challenges have been solved in several countries.

# 8  Three Framing Strategies: From a Complex Reality to a Categorical Label

65. There was indeed a discussion of this last point addressed by KRAV – the organization that ran the Swedish seafood labelling project – but after some scornful comments from stakeholders this suggestion was quickly withdrawn (Boström, 2004b).
66. The order in which these three strategies are presented here does not reflect any ideal or real order of framing processes in general or of labelling in particular.
67. The definition of frame bridging provided by Snow and colleagues (1986, p. 467) is useful: 'By frame bridging we refer to the linkage of two or more ideologically congruent but structurally unconnected frames regarding a particular issue or problem.'
68. Frame extension is the strategy of actors to extend their frames beyond their initial interests, goals, and knowledge basis so as to increase frame resonance (Snow et al., 1986; Snow & Benford, 1988; cf. Gamson, 1992 on 'cultural resonance').
69. John A. Fagan, Science-Based, Precautionary Engineered Foods (2000); available online at http://www.geocities.com/luizmeira/label.html (accessed 3 March 2001). John Fagan is Professor of Molecular Biology at Maharishi University of Management in Iowa.
70. FDA (Food and Drug Administration) (1992) Statement of policy: foods derived from new plant varieties. Federal Register, 57, 22984–23005 (WWW document). URL http://vm.cfsan.fda.gov/~lrd/biocon.html (accessed 18 July 2006).
71. See, for example, Klintman & Boström (2004) for more details of this particular debate.
72. Metaframing differs from boundary framing in that the latter has its focus on the boundary between the desired and undesired components. Metaframing, in contrast, is a strategy of including components across boundaries. Thus, metaframing also differs from frame bridging, where various groups and positions across which a 'bridge' is framed (e.g., environmental movement and everyday consumer motives) do not appear to be polarized, but ideologically congruent.
73. 'Sustainability' and 'ecomodernism' could very well be understood as metaframes in that they have been developed by combining (and transforming) opposite discourses on 'economic growth' and 'limits to growth'.
74. For instance, large retailers in Sweden (e.g., Hemköp) have begun to favour 'ecological' Christmas ham, which includes nitrite, rather than the

KRAV-labelled ham, which does not include nitrite. Retailers claim that consumers favour the ham with nitrite since it gives the meat its pink colour (13 December 2004, http://www2.unt.se/avd/1,1786,MC=7-AV_ID=367191,00.html, accessed 16 July 2007).

75. A positive note, however, is that KRAV is currently engaged in developing climate labelling for food products. The idea is not to include it in the organic framework but to develop an independent system. The future will show how successful the attempt will be.

# 9   Organizing the Labelling

76. The Swedish seafood labelling case is a good example. Swedish fishermen's associations and related industries had expressed a hostile attitude to the WWF-led MSC initiative. However, when fishermen came under severe media and public attack in the years around the turn of the millennium, many business actors gradually understood they had to do something to regain trustworthiness. They developed a friendlier attitude towards EMOs such as the WWF and gradually business actors committed to the discussions about introducing an eco-labelling system, in the hope that such a system would create credibility and good PR for the business.

77. Such activities include the organizing of policy development, standards development, standards interpretation, fund-raising, and marketing activities. They may furthermore include activities such as accreditation of certification bodies, facilitation of communication with consumers, monitoring of performance of licence holders and certification bodies, and prevention of misused labels.

78. Nevertheless, issues such as transparency, public involvement, and accountability are often standard criteria for companies to be 'labelled' as ethical or socially responsible, not least under green and ethical mutual funds.

79. The Domini Social 400 Index is the exclusive property of the KLD Research & Analytics, Inc.

80. Likewise, the European FSC-competing model PEFC – which has eventually become a global model of which SFI is a member – mirrors the FSC's tripartite structure. The members of PEFC, as are those of the FSC, are divided into three equal groups: (1) forestry, (2) primary processing industry, and (3) other interests. In this form, the business side (1 and 2) gets a majority position, which many EMOs do not accept. The formal structure allows for the participation of EMOs, but the main EMOs such as WWF and FoE have chosen not to participate.

81. To use KRAV as an example, it is stated in the constitution that a member must be 'a national association, another association or a single company with a significant position within its industry' (see Boström, 2006a). All member organizations have one vote in the annual KRAV assembly, which is the highest forum for decision-making. The board must, at a minimum, consist of two representatives from agriculture (including at least one from organic agriculture); two from trade (retailers); one from processing industries; and two from consumer, environmental, and animal-welfare groups. Because decision-making follows the majority principle, the business side

could, theoretically, achieve dominance over SMOs on the board (if the organic party and the consumer groups, environmental and animal-welfare groups are seen as SMOs).

82. Gale makes a systematic comparison of the development of regional FSC standards in Canada and the United States. He notes that 'more bottom-up negotiation arrangements are associated with more demanding forest management standards, the more top-down with less' (2004, p. 80), which is a finding that supports our argument here. The bottom-up approach implies the inclusion of more SMO-type actors.

83. Additional differences concerned, for example, auditing and chain-of-custody arrangements (see also Domask, 2003). Initially, the SFI programme did not have to be independently audited by external organizations, as members could either audit themselves (first-party auditing) or have the AF&PA do it (second-party auditing). The SFI has been reluctant to develop a stringent chain-of-custody arrangement, which indeed is essential for the possibility of tracking labelled products to certified raw material.

84. For example, in 1998 the SFI changed its policy to allow third-party auditing and it addressed chain-of custody issues (Cashore et al., 2004). The environmental groups that participate within SFI 'aggressively pursue more strict environmental standards or threaten that they will have to resign to protect their image' (Cubbage & Newman, 2005, p. 266). The FSC, for its part, has tried over time to develop more flexible mechanisms in line with market-pragmatic thinking, especially in the United States, because FSC promoters had to face the potential disappearance of the FSC as a certification programme in the US context (Cashore et al., 2004). 'FSC has become more pragmatic in their operations, especially in implementation, if not on paper' (Cubbage & Newman, 2005, p. 266). Likewise, the competitor programmes have adopted certain substantive rules. It has been claimed that PEFC in certain parts of Sweden is more similar to the FSC in Sweden than is PEFC in other regions (where the FSC is lacking) (see Lindahl, 2001).

85. For instance, the American thresholds for the label 'Made with organic ingredients' can be used for products with 70–95 per cent organic content; this was partially an adaptation to the thresholds used in the European Union. The fact that 70 per cent was the lower limit in the EU in the early 2000s was partially something that motivated the National Organic Program to raise the bar, for reasons of international trade (Klintman, 2002b).

86. Although the labelling process was administrated by KRAV, which is a body in which SMOs are represented in decision-making (see above), this specific project used a specific organizational form. The reason behind the choice was that the project was extraordinarily controversial, with huge mutual mistrust between fishing industries and EMOs (see Chapter 10), so it was not possible to include the latter group as decision-makers (Boström, 2006b). The KRAV staff believed they had to design a particular organizational structure, which was biased to the advantage of fishing industries (representatives of the fishermen, fish processors, retailers, professional buyers, and marine research).

87. Södra Dalarnes Tidning, 16 August 2003; interview with Johan Kling, an expert on electricity transportation at the Swedish Society for Nature Conservation (SSNC), 14 September 2004. Moreover, SSNC, which also used

to be a frontrunner in eco-labelling of paper products, is currently finding that not a single company wants to use Good Environmental Choice for most types of paper products (Nilsson, 2005, pp. 14–15). SSNC ended up making its model irrelevant to the paper market, because of failure to meet market-pragmatic goals.

88. See http://www.envocare.co.uk/ethical_investment_criteria.htm (accessed 25 October 2007).

89. The ISO 14000 is considered the most widely recognized global-level voluntary initiative on the part of the industry (Clapp, 2005). By 2001, almost 50,000 firms in 118 countries had gained ISO 14001 certification (ibid.).

90. The fact that the FSC's very democratic structure has caused many protracted conflicts and debates is confirmed and reported by Timothy Synnott, a previous Executive Director of the FSC and one of the key figures in the FSC's establishment, in his notes on the early years of the FSC (Synnott, 2005). The first General Assembly in 1996 almost caused a meltdown, because of the strong differences of opinions among FSC members, he recalls.

91. Benjamin Caspar, team coordinator, http://ec.europa.eu/environment/ecolabel/pdf/marketing/management_group/minutes_mmg_retailer_040406.pdf, accessed 17 July 2007.

92. http://www.domini.com/about-domini/The-Domini-Story/index.htm, accessed 25 October 2007.

93. One moderate variant of coalition-building after the establishment of a labelling organization is the building of buyers' networks, such as WWF's Global Forest and Trade Network. The idea is to organize a network to visualize a demand for FSC-certified raw material, and a joint pressure in the face of forest producers.

# 10   Dealing with Mutual Mistrust

94. An inclusive labelling organization provides a setting for repeated dialogue and negotiation surrounding labelling policies and standards specifically, but possibly also concerning green consumerist policies in general. Other scholars have observed that repeated interaction over time in organized networks comprising a wide array of actors can result in mutual learning, mutual trust, and common expectations of proper behaviour (Cutler et al., 1999; Sabatier & Jenkins-Smith, 1999; Rhodes, 2000; Elliot & Schlaepfer, 2001; Wälti et al., 2004).

95. It could, however, be seen as systematic in the sense that science is reflectively (or systematically) assessed from a firm framework which includes notions of naturalness.

# 11   Green Labelling and Green Consumerism: Challenges and Horizons

96. http://www.green-e.org/, accessed 12 October 2007.

97. An interesting review of these aspects can be found in a book chapter by Karl and Orwat (1999). They suggest that increasing competition among

labels will require tighter criteria and may therefore help to alleviate many of the problems generated by labelling schemes and increase the overall credibility of the labels. Competition may also create confusion among consumers, however, which runs counter to one of the main motives behind the creation of eco-labels. Therefore various institutions, such as research and test centres, are needed to support consumer decisions.

# References

Adams, C. and Zutshi, A. (2005) 'Corporate Disclosure and Auditing' in R. Harrison, T. Newholm and D. Shaw (eds) *The Ethical Consumer* (London: Sage).

Ahrne, G. (1994) *Social Organizations. Interaction Inside, Outside and Between Organizations* (London: Sage).

Ahrne, G. and Brunsson, N. (eds) (2004a). *Regelexplosionen* (Stockholm: EFI).

Ahrne, G. and Brunsson, N. (eds) (2004b) 'Soft Regulation from an Organizational Perspective', in U. Mörth (ed.) *Soft Law in Governance and Regulation* (Cheltenham: Edward Elgar).

Ahrne, G. and Brunsson, N. (2008) *Meta-organizations* (Cheltenham: Edward Elgar).

Allen, P. and Covack, M. (2000) 'The Capitalist Composition of Organic: The Potential of Markets in Fulfilling the Promise of Organic Agriculture', *Agriculture and Human Values*, 17, 221–232.

Alternative Farming Systems Information Center (2001) *Organic Food Production*. USDA. Available online: http://www.nalusda.gov/afsic/ofp/ofp.htm, accessed 16 November 2001.

Amaditz, K. C. (1997) 'The Organic Foods Production Act of 1990 and its Impending Regulations: A Big Zero for Organic Food?', *Food Drug Law Journal*, 52(4), 537–559.

Barham, E. (2002) 'Towards a Theory of Values Based Labelling', *Agriculture and Human Values*, 19, 349–360.

Batte, M. T. Hooker, N. H., Haab, T. C. and Beaverson, J. (2007) 'Putting their Money where their Mouths are: Consumer Willingness to Pay for Multi-ingredient, Processed Organic Food Products', *Food Policy*, 32(2), 145–159.

Bauman, Z. (1991) *Modernity and Ambivalence* (Cambridge: Polity Press).

Beabout, G. R. and Schmiesing, K. E. (2003) 'Socially Responsible Investing: An Application of Catholic Social Thought', *Logos*, 6, 63–99.

Beck, U. (1986/1992) *Risk Society. Towards a New Modernity* (London: Sage Publications).

Beck, U. (1994) 'The Reinvention of Politics: Towards a Theory of Reflexive Modernization', pp. 1–55 in U. Beck, A. Giddens and S. Lash (eds) *Reflexive Modernization. Politics, Tradition and Aesthetics in the Modern Social Order* (Cambridge: Polity Press).

Beck, U. (2000) *What is Globalization?* (Cambridge: Polity Press).

Bendell, J. (ed.) (2000) *Terms for Endearment: Business, NGOs and Sustainable Development* (Sheffield: Greenleaf Publishing).

Bendell, J. and Murphy, D. (2000) 'Planting the Seeds of Change. Business-NGO Relations on Tropical Deforestation', pp. 65–78 in J. Bendell (ed.) *Terms for Endearment. Business, NGOs and Sustainable Development* (Sheffield, UK: Greenleaf Publishing).

Benford, R. D. and Snow, D. (2000) 'Framing Processes and Social Movements: An Overview and Assessment', *Annual Review of Sociology*, 26, 611–639.

Bernstein, S. and Cashore, B. (2004) 'Non-State Global Governance: Is Forest Certification a Legitimate Alternative to a Global Forest Convention?', pp. 33–65 in J. Kirton and M. Trebilcock (eds) *Hard Choices, Soft Law: Combining Trade, Environment, and Social Cohesion in Global Governance* (Aldershot: Ashgate Press).

Berry, H. and McEachern, M. (2005) 'Informing Ethical Consumers', in R. Harrison, T. Newholm and D. Shaw (eds) *The Ethical Consumer* (London: Sage).

Bertoldi, P. (1999) 'Energy Efficient Equipment within SAVE: Activities, Strategies, Success and Barriers', Proceedings of the SAVE Conference for an Energy Efficient Millennium, Graz, Austria, 1999.

Bhaskar, R. (1989) *Reclaiming Reality: A Critical Introduction to Contemporary Philosophy* (London: Verso).

Blyth, M. (2002) *Great Transformations. Economic Ideas and Institutional Change in the Twentieth Century* (Cambridge, UK: Cambridge University Press).

Boli, J. and Thomas, G. (1999) 'INGOs and the Organization of World Culture', pp. 13–49 in J. Boli and G. Thomas (eds) *Constructing World Culture: International Nongovernmental Organizations since 1875* (Stanford: Stanford University Press).

Boström, M. (1999) 'Den Organiserade Miljörörelsen. Fallstudier av Svenska Naturskyddsföreningen, Världsnaturfonden WWF, Miljöförbundet Jordens Vänner, Greenpeace och Det Naturliga Steget', *Score Working Paper Series 1999:9* (Stockholm: Score).

Boström, M. (2001) *Miljörörelsens Mångfald* (Lund: Arkiv).

Boström, M. (2002) 'Skogen Märks – Hur Svensk Skogscertifiering kom till och dess Konsekvenser', *Score Working Paper Series 2002:3* (Stockholm: Score). Electronically available at www.score.su.se.

Boström, M. (2003a) 'Environmental Organizations in New Forms of Political Participation: Ecological Modernization and the Making of Voluntary Rules', *Environmental Values*, 12, 175–193.

Boström, M. (2003b) 'How State-dependent is a Nonstate-driven Rule-making Project? The Case of Forest Certification in Sweden', *Journal of Environmental Policy & Planning*, 5, 165–180.

Boström, M. (2004a) 'Cognitive Practices and Collective Identities within a Heterogeneous Social Movement: The Swedish Environmental Movement', *Social Movement Studies*, 3, 73–88.

Boström, M. (2004b) 'Fina Fisken – En Studie av Bakgrunden, Organiseringen och Debatten kring Introduktionen av Miljömärkt Vildfångad Fisk i Sverige', *Score Working Paper Series 2004:9* (Stockholm: Score). Electronically available at www.score.su.se.

Boström, M. (2006a) 'Establishing Credibility: Practising Standard-setting Ideals in a Swedish Seafood-labelling Case', *Journal of Environmental Policy & Planning*, 8, 135–158.

Boström, M. (2006b) 'Regulatory Credibility and Authority through Inclusiveness. Standardization Organizations in Cases of Eco-labelling', *Organization*, 13, 345–367.

Boström, M. (2007) 'The Historical and Contemporary Roles of Nature Protection Organisations in Sweden', in C. S. A. van Koppen and W. T. Markham (eds)

*Protecting Nature: Organizations and Networks in Europe and the U.S.* (Cheltenham: Edward Elgar).

Boström, M. and Klintman, M. (2003). 'Framing, Debating, and Standardizing "Natural Food" in Two Different Political Contexts', *Score Working Paper Series 2003:3* (Stockholm: Score). Electronically available at www.score.su.se.

Boström, M. and Klintman, M. (2006a) 'State-Centred versus Non-State-Driven Organic Food Standardization: A Comparison of the U.S. and Sweden', *Agriculture and Human Values*, 23, 163–180.

Boström, M. and Klintman, M. (2006b) 'Hur Översätts och Förhandlas Komplex Kunskap till ett Kategoriskt Miljömärke?', pp. 79–108 in K. Fernler and C.-F. Helgesson (eds) *Kloka Regler? Kunskap i Regelsamhället* (Lund: Studentlitteratur).

Boström, M. and Garsten, C. (2008) (eds) *Organizing Transnational Accountability* (Cheltenham: Edward Elgar).

Boström, M., Føllesdal, A., Klintman, M., Micheletti, M. and Sørensen, M.P. (eds) (2005) *Political Consumerism: Its Motivations, Power, and Conditions in the Nordic Countries and Elsewhere* (Copenhagen: Nordic Council of Ministers). Electronically available at http://www.norden.org/pub/velfaerd/konsument/sk/TN2005517.pdf.

Boulanger, P.-M. and Zaccai, E. (2007) 'Conclusions: the Future of Sustainable Consumption' pp. 231–238 in E. Zaccai (ed.) *Sustainable Consumption, Ecology and Fair Trade* (London: Routledge).

Bowker, G. C. and Leigh Star, S. (1999) *Sorting Things Out. Classification and Its Consequences* (Cambridge, MA: MIT Press).

Brunsson, N. and Jacobsson, B. (eds) (2000) *A World of Standards* (Oxford: Oxford University Press).

Brunsson, N. and Sahlin-Andersson, K. (2000) 'Constructing Organizations. The Example of Public Sector Reform', *Organization Studies*, 21, 721–746.

van den Burg, S. (2006) *Governance through Information*. PhD thesis (Wageningen: Wageningen University).

van den Burg, S. and Mol, A. P. J. (2008) 'Make It All Publicly Available: Four Challenges to Environmental Disclosure' in M. Boström and C. Garsten (eds) *Organizing Transnational Accountability* (Cheltenham: Edward Elgar).

Busch, L. (2000). 'The Moral Economy of Grades and Standards', *Journal of Rural Studies*, 16, 273–283.

Carson, M. (2004) *From Common Market to Social Europe? Paradigm Shift and Institutional Change* (Stockholm: Almqvist & Wicksell International).

Cashore, B., Auld, G. and Newsom, D. (2004) *Governing through Markets: Forest Certification and the Emergence of Non-State Authority* (New Haven: Yale University Press).

Cheftel, J. C. ( 2005) 'Food and Nutrition Labelling in the European Union', *Food Chemistry*, 93, 531–550.

Chong, D. and Druckman, J. N. (2007) 'Framing Theory', *Annual Review of Political Science*, 10, 103–126.

Christensen, T. and Peters, G. (1999) *Structure, Culture, and Governance: A Comparison of Norway and the United States* (Oxford: Rowman & Littlefield Publishers).

Clapp, J. (2005) 'Transnational Corporations and Global Environmental Governance', pp. 284–297 in P. Dauvergne (ed.) *Handbook of Global Environmental Politics* (Northampton: Edward Elgar).

Cochoy, F. (2004) 'The Industrial Roots of Contemporary Political Consumerism: the Case of the French Standardization Movement', pp. 145–160 in M. Micheletti, A. Follesdal and D. Stolle (eds) *Politics, Products, and Markets. Exploring Political Consumerism Past and Present* (New Brunswick, London: Transaction Publishers).

Coetzee, J. M. (2003) *Youth* (London: Vintage).

Cohen, L. (2003) *A Consumers' Republic: The Politics of Mass Consumption in Postwar America* (NYC: Vintage).

Cohen, M. (2005) Sustainable Consumption and Global Citizenship: An Empirical Analysis', in M. Boström, A. Føllesdal, M. Klintman, M. Micheletti and M. P. Sørensen (eds) *Political Consumerism: Its Motivations, Power, and Conditions in the Nordic Countries and Elsewhere* (Copenhagen: Nordic Council of Ministers). Electronically available at http://www.norden.org/pub/velfaerd/konsument/sk/TN2005517.pdf.

Cohen, M. and Murphy, J. (2001) *Exploring Sustainable Consumption. Environmental Policy and the Social Sciences* (Oxford: Elsevier).

Cohen, M., Comrov, A. and Hoffner, B. (2005) 'The New Politics of Consumption: Promoting Sustainability in the American Marketplace', *Sustainability: Science, Practice, & Policy*, 1, 58–76.

Collins, H. and Pinch, T. (1993) *The Golem. What Everyone Should Know about Science* (Cambridge: Cambridge University Press).

Connelly, J. and Smith, G. (2003) *Politics and the Environment. From Theory to Practice* (2nd edition) (London: Routledge).

Constance, D. and Bonanno, A. (2000) 'Regulating the Global Fisheries: The World Wildlife Fund, Unilever, and the Marine Stewardship Council', *Agriculture and Human Values*, 17, 125–139.

Crane, A. (2000) 'Facing the Backlash: Green Marketing and Strategic Reorientation in the 1990s', *Journal of Strategic Marketing*, 8, 277–296.

Crane, A. (2005) 'Meeting the Ethical Gaze: Challenges for Orienting to the Ethical Market', in R. Harrison, T. Newholm and D. Shaw (eds) *The Ethical Consumer* (London: Sage).

Cubbage, F. W. and Newman, D. H. (2006) 'Forest Policy Reformed: A United States Perspective', *Forest Policy and Economics*, 9, 262–273.

Cutler, C., Haufler, V. and Porter, T. (eds) (1999) *Private Authority and International Affairs* (Albany: State University of New York Press).

Darby, M. and Karni, E. (1973) 'Free Competition and Optimal Amount of Fraud', *Journal of Law and Economics*, 16, 67–88.

Deere, C. (1999) *Eco-labelling and Sustainable Fisheries* (Washington DC: IUCN and Rom: FAO).

De Pelsmacker, P. and Janssens, W. (2007) 'A Model for Fair Trade Buying Behaviour: The Role of Perceived Quantity and Quality of Information and of Product-specific Attitudes', *Journal of Business Ethics*, 75, 361–380.

Del Río, P. and Gual, M. A. (2004) 'The Promotion of Green Electricity in Europe: Present and Future', *European Environment*, 14, 219–234.

Dingwerth, K. (2005) *The Democratic Legitimacy of Transnational Rule-Making: Normative Theory and Democratic Practice*. PhD dissertation (Berlin: Freie Universität Berlin).

Djelic, M.-L. and Sahlin-Andersson, K. (2006) (eds) *Transnational Governance. Institutional Dynamics of Regulation* (Cambridge: Cambridge University Press).

216   *References*

Domask, J. (2003) 'From Boycotts to Global Partnership: NGOs, the Private Sector, and the Struggle to Protect the World's Forests', in J. Doh and H. Teegen (eds) *Globalization and NGOs. Transforming Business, Government, and Society* (London: Praeger).

Domini (2007) 'The Domini Story', electronically available at http://www.domini.com/about-domini/The-domini-story/index.htm, accessed 14 June 2008.

Dryzek, J. S. (1993) 'Policy Analysis and Planning: From Science to Argument', pp. 214–232 in F. Fischer and J. Forester (eds) *The Argumentative Turn in Policy Analysis* (Durham, NC: Duke University Press).

Dryzek, J. S. (2001) 'Legitimacy and Economy in Deliberative Democracy', *Political Theory*, 29, 651–669.

Dryzek, J. S., Downes, D. and Hunold, C. (2003) *Green States and Social Movements: Environmentalism in the United States, Britain, Germany, and Norway* (Oxford: Oxford University Press).

Eder, K. (1996) *The Social Construction of Nature* (London: Sage Publications).

Egan, M. (2001) *Constructing a European Market: Standards, Regulation, and Governance* (Oxford: Oxford University Press).

Ek, K. (2005) 'Public and Private Attitudes Towards "Green" Electricity: The Case of Swedish Wind Power', *Energy Policy*, 33, 1677–1689.

Ekelund (Axelson), L. (2003) 'På spaning efter den ekologiska konsumenten: En genomgång av 25 svenska konsumentundersökningar på livsmedelsområdet. Ekologiskt lantbruk 39' (Centrum för uthålligt lantbruk, SLU, Uppsala).

Ekelund, L. and Tjärnemo, H. (2004) 'Consumer Preferences for Organic Vegetables – The Case of Sweden', *Acta Horticulturae*, 655, 121–128.

Elad, C. (2001) 'Auditing and Governance in the Forestry Industry: Between Protest and Professionalism', *Critical Perspectives on Accounting*, 12, 647–671.

Elkington, J., Hailes, J. and Makower, J. (1990) *The Green Consumer Guide* (London: Penguin).

Elliot, C. (1999) *Forest Certification: Analysis From a Policy Network Perspective*. PhD dissertation (Lausanne: Département de génie rural, Ecole Polytechnique Fédérale de Lausanne).

Elliot, C. and Schlaepfer, R. (2001) 'The Advocacy Coalition Framework: Application to the Policy Process for the Development of Forest Certification in Sweden', *Journal of European Public Policy*, 8, 642–661.

Erskine, C. C. and Collins, L. (1997) 'Eco-labelling: Success or Failure?' *The Environmentalist*, 17, 125–133.

Europe, history of (2008), in 'Encyclopædia Britannica', electronically available at http://www.britannica.com/EBchecked/topic/195896/history-of-Europe, accessed 7 July 2008.

Eyerman, R. and Jamison, A. (1991) *Social Movements. A Cognitive Approach* (Cambridge: Polity Press).

Fagan, J. A. (2000) 'Science-Based, Precautionary Engineered Foods', http://www.geocities.com/luizmeira/label.html, accessed 3 March 2001.

Fernau, K. (2001) 'Going organic becomes simple', *Arizona Republic*, 31 January 2001, at G3.

Fischer, F. (2003) *Reframing Public Policy* (Oxford: Oxford University Press).

FoodFirst (2000) Let Organic – Stay Organic, 31 January, Oakland, CA: FoodFirst/Institute for Food and Development Policy, http://www.foodfirst.org/action/2000/1–31–00.html, accessed 14 November 2000.

Fowler, P. and Heap, S. (2000) 'Bridging Troubled Waters: The Marine Stewardship Council', pp. 135–149 in J. Bendell (ed.) *Terms for Endearment. Business, NGOs and Sustainable Development* (Sheffield, Greenleaf Publishing).

Frankel, G. C. and Borque, M. (1998) *Certified Organic: Recent Developments in the Proposed Rule For a National "Organic" Standard in the U.S.* Electronically available at http://www.foodfirst.org/progs/global/susag/.

Friedman, A. L. and Miles, S. (2001) 'Socially responsible investment and corporate social and environmental reporting in the UK: An exploratory study', *The British Accounting Review*, 33(4), 523–548.

Friedman, M. (1999) *Consumer Boycotts: Effecting Change through the Marketplace and the Media* (London: Routledge).

Gale, F. (2004) 'The Consultation Dilemma in Private Regulatory Regimes: Negotiating FSC Regional Standards in the United States and Canada', *Journal of Environmental Policy and Planning*, 6, 57–84.

Gallastegui, I. G. (2002) 'The Use of Eco-labels: a Review of the Literature', *European Environment*, 12, 316–331.

Gan, L., Eskeland, G. and Kolshus, H. (2007) 'Green Electricity Market Development. Lessons from Europe and the US', *Energy Policy*, 35, 144–155.

Garsten, C. (2008) 'The United Nations – Soft and Hard: Regulating Social Accountability for Global Business', in M. Boström and C. Garsten (eds) *Organizing Transnational Accountability* (Cheltenham: Edward Elgar).

Getz, C. and Shreck, A. (2006) 'What Organic and Fair Trade Labels do not Tell us: Towards a Place-based Understanding of Certification', *International Journal of Consumer Studies*, 30, 490–501.

Giddens, A. (1984) *The Constitution of Society. Outline of the Theory of Structuration* (Berkeley: University of California Press).

Giddens, A. (1990) *The Consequences of Modernity* (Cambridge: Polity Press).

Giddens, A. (1991) *Modernity and Self-Identity* (Stanford CA: Stanford University Press).

Gieryn, T. (1983) 'Boundary-work and the Demarcation of Science from Non-science', *American Sociological Review*, 48, 781–795.

Gilg, A., Barr, S. and Ford, N. (2005) 'Green Consumption or Sustainable Lifestyles? Identifying the Sustainable Consumer', *Futures*, 37, 481–504.

Glasbergen, P., Biermann, F. and Mol, A. P. J. (eds) (2007) *Partnerships, Governance and Sustainable Development. Reflections on Theory and Practice* (Cheltenham: Edward Elgar).

Goffman, E. (1974) *Frame Analysis: An Essay on the Organization of Experience* (New York: Harper & Row).

Golan, E., Kuchler, F. and Mitchell, L. (2000) *Economics of Food Labelling* (Washington DC: U.S. Dept. of Agriculture, 793), http://www.ers.usda.gov/publications/aer793/, accessed 4 January 2008.

Goldemberg, J. (2006) 'The Promise of Clean Energy', *Energy Policy*, 34, 2185–2190.

Goul Andersen, J. and Tobiasen, M. (2004) 'Who Are These Political Consumers Anyway? Survey Evidence from Denmark', in M. Micheletti, A. Follesdal and D. Stolle (eds) *Politics, Products, and Markets. Exploring Political Consumerism Past and Present* (New Brunswick, London: Transaction Publishers).

Grankvist, G., Dahlstrand, U. and Biel, A. (2004) 'The Impact of Environmental Labelling on Consumer Preference: Negative vs. Positive Labels', *Journal of Consumer Policy*, 27, 213–230.

Green, K., Morton, B. and New, S. (2000) 'Greening Organizations, Purchasing, Consumption, and Innovation', *Organization and Environment*, 13, 206–225.

Greenberg, C. (2004) 'Political Consumer Action: Some Cautionary Notes from African American History', pp. 63–82 in M. Micheletti, A. Follesdal and D. Stolle (eds) *Politics, Products, and Markets. Exploring Political Consumerism Past and Present* (New Brunswick, London: Transaction Publishers).

Gregory, R. (2003) 'Accountability in Modern Government', pp. 557–568 in J. Pierre and G. Peters (eds) *Handbook of Public Administration* (London: Sage).

Gulbrandsen, L. H. (2004) 'Overlapping Public and Private Governance: Can Forest Certification Fill the Gaps in the Global Forest Regime?', *Global Environmental Politics*, 4, 75–99.

Gulbrandsen, L. H. (2005a) 'Explaining Different Approaches to Voluntary Standards: a Study of Forest Certification Choices in Norway and Sweden', *Journal of Environmental Policy and Planning*, 7, 43–59.

Gulbrandsen, L. H. (2005b) 'Mark of Sustainability? Challenges for Fishery and Forestry Eco-labelling', *Environment*, 47, 8–23.

Gulbrandsen, L. H. (2006) 'Creating Markets for Eco-labelling: are Consumers Insignificant?', *International Journal of Consumer Studies*, 30, 477–489.

Gulbrandsen, L. H. (2008) 'Organizing Accountability in Transnational Standards Organizations: the Forest Stewardship Council as a Good Governance Model', in M. Boström and C. Garsten (eds) *Organizing Transnational Accountability* (Cheltenham: Edward Elgar).

Guthman, J. (2004) *Agrarian Dreams: The Paradox of Organic Farming in California* (Los Angeles: California University Press).

Hajer, M. (1995) *The Politics of Environmental Discourse. Ecological Modernisation and the Policy Process* (Oxford: Clarendon Press).

Halkier, B. (2001) 'Consuming Ambivalences: Consumer Handling of Environmentally Related Risks in Food', *Journal of Consumer Culture*, 1, 205–224.

Halkier, B. (2004) 'Consumption, Risk, and Civic Engagement: Citizens as Risk-Handlers', pp. 223–244 in M. Micheletti, A. Follesdal and D. Stolle (eds) *Politics, Products, and Markets. Exploring Political Consumerism Past and Present* (New Brunswick, London: Transaction Publishers).

Hall, R. B. and Biersteker, T. (eds) (2002) *The Emergence of Private Authority in Global Governance* (Cambridge: Cambridge University Press).

Hancher, L. and Moran, M. (eds) (1989) *Capitalism, Culture, and Economic Regulation* (Oxford: Clarendon Press).

Hannigan, J. (2006) *Environmental Sociology* (2nd edition) (London: Routledge).

Hardin, R. (2006) *Trust* (Cambridge: Polity Press).

Harrison, R., Newholm, T. and Shaw, D. (eds) (2005) *The Ethical Consumer* (London: Sage).

Hasselberg, Y. (1997) 'Mål och Makt i Svensk Fiskeripolitik', in L. Hultkrantz, Y. Hasselberg and D. Stigberg (eds) *Fisk och Fusk – Mål, Medel och Makt i fiskeripolitiken* (Finansdepartementet DS 1997:81).

Held, D., McGrew, A., Goldblatt, D. and Perraton, J. (1999) *Global Transformations* (Stanford: Stanford University Press).

Hill, R. P., Ainscough, T., Shank, T. and Manullang, D. (2007). 'Corporate Social Responsibility and Socially Responsible Investing: A Global Perspective', *Journal of Business Ethics*, 70, 165–174.

Hofer, K. (2000). 'Labelling of organic food products', in A. Mol, V. Lauber, and D. Liefferink (eds) *The Voluntary Approach to Environmental Policy: Joint Environmental Policy-making in Europe* (Oxford: Oxford University Press).

Holzer, B. (2006) 'Political Consumerism Between Individual Choice and Collective Action: Social Movements, Role Mobilization and Signalling', *International Journal of Consumer Studies*, 30, 405–415.

Holzer, B. (2007) 'Framing the Corporation: Royal Dutch/Shell and Human Rights in Nigeria', *Journal of Consumer Policy*, 30, 281–301.

Howes, R. (2005) 'Reversing the decline in global fish stocks: Eco-labelling and the Marine Stewardship Council', *Sustainable Development International*, 13 (spring edition), http://www.sustdev.org/index.php?option=com_content &task=view&id=365&Itemid=34, accessed 18 August 2005.

Hultkrantz, L., Hasselberg, Y. and Stigberg, D. (1997) *Fisk och Fusk – Mål, Medel och Makt i Fiskeripolitiken* (Finansdepartementet DS 1997:81).

Humphrey, C. and Owen, D. (2000) 'Debating the "Power" of audit', *International Journal of Auditing*, 4, 29–50.

Hunt, S. A., Benford, R. D. and Snow, D. A. (1994) 'Identity fields', pp. 185–208 in E. Larana, H. Johnston, and J. Gusfield (eds) *New Social Movements: From Ideology to Identity* (Philadelphia: Temple University Press).

Höijer, B., Lidskog, R. and Uggla, Y. (2005) 'Facing Dilemmas: Sense-making and Decision-making in Late Modernity', *Futures*, 38, 350–366.

Irwin, A. and Wynne, B. (eds) (1996) *Misunderstanding Science? The Public Reconstruction of Science and Technology* (Cambridge: Cambridge University Press).

Jacobsson, B. (2000) 'Standardization and Expert Knowledge', in N. Brunsson and B. Jacobsson (eds) *A World of Standards* (Oxford: Oxford University Press).

Jamison, A., Eyerman, R. and Cramer, J. (1990) *The Making of the New Environmental Consciousness. A Comparative Study of the Environmental Movements in Sweden, Denmark and the Netherlands* (Edinburgh: Edinburgh University Press).

Jasanoff, S. (2005) *Designs on Nature. Science and Democracy in Europe and the United States* (Princeton: Princeton University Press).

Johansson, B. (ed.) (2003) *Torskar Torsken? Forskare och Fiskare om Fisk och Fiske* (Stockholm: Formas).

Jordan, A., Wurzel, R. and Zito, A. (guest eds) (2003) ' "New" Instruments of Environmental Governance: National Experiences and Prospects', *Environmental Politics*, 12, 1–24.

Jordan, A., Wurzel, R., Zito, A. and Brückner, L. (2004) 'Consumer Responsibility-taking and Eco-labelling Schemes in Europe', pp. 161–180 in M. Micheletti, A. Follesdal and D. Stolle (eds) *Politics, Products, and Markets. Exploring Political Consumerism Past and Present* (New Brunswick, London: Transaction Publishers).

Karl, H. and Orwat, C. (1999) 'Environmental Labelling in Europe: European and National Tasks', *European Environment*, 9, 212–220.

Kerwer, D. (2008) 'Watchdogs Beyond Control? The Accountability of Accounting Standards Organizations', in M. Boström and C. Garsten (eds) *Organizing Transnational Accountability* (Cheltenham: Edward Elgar).

Kitschelt, H. P. (1986) 'Political Opportunity Structures and Political Protest: Anti-nuclear Movements in Four Democracies', *British Journal of Political Science*, 16, 57–85.

Kjærnes, U., Harvey, M. and Warde, A. (2007) *Trust in Food: A Comparative and Institutional Analysis* (London/Basingstoke, UK: Palgrave Macmillan).

Klintman, M. (2000) *Nature and the Social Sciences: Examples from the Electricity and Waste Sectors* (Lund: Lund Dissertations in Sociology).

Klintman, M. (2002a) 'Arguments Surrounding Organic and Genetically Modified Food Labelling: A Few Comparisons', *Journal of Environmental Policy & Planning*, 4, 247–259.

Klintman, M. (2002b) 'The Genetically Modified (GM) Food Labelling Controversy: Ideological and Epistemic Crossovers', *Social Studies of Science*, 32(1), 71–91.

Klintman, M. (2006) 'Ambiguous Framings of Political Consumerism: Means or End, Product or Process Orientation?', *International Journal of Consumer Studies*, 30, 427–438.

Klintman, M. and Boström, M. (2004). 'Framings of Science and Ideology: Organic Food Labelling in the US and Sweden', *Environmental Politics*, 13(3), 612–633.

Klintman, M. and Boström, M. (guest eds) (2006) 'Political and Ethical Consumerism Around the World', *International Journal of Consumer Studies*, 30 (5).

Klintman, M. and Boström, M. (2008) 'Transparency Through Labelling? Layers of Visibility in Environmental Risk Management', in C. Garsten and M. Lindh de Montoya (eds) *Transparency in a New Global Order: Unveiling Organizational Visions* (Cheltenham: Edward Elgar).

Klintman, M., Boström, M., Ekelund, L. and Lindén, A-L. (2008) *Maten Märks* (Lund: Lund University, Dept. of Sociology, Research Reports in Sociology).

Klintman, M., Mårtenson, K. with Johansson, M. (2003), 'Bioenergi för Uppvärmning – Hushållens Perspektiv', Research Report in Sociology, 2003:1, Lund University. Electronically available at http://www.fpi.lu.se/en/klintman.

Klonsky, K. (2000) 'Forces Impacting the Production of Organic Foods', *Agriculture and Human Values*, 17, 233–243.

van Koppen, C. S. A. and Markham, W. T. (eds) (2007) *Protecting Nature: Organizations and Networks in Europe and the U.S.* (Cheltenham: Edward Elgar).

KRAV (2000) *15 Goda år med KRAV: Vi Skriver Historia* (Uppsala: KRAV).

Krumsick, B. (2003) 'Socially Responsible High Tech Companies: Emerging Issues', *Journal of Business Ethics*, 43, 179–187.

Lafferty, W. and Meadowcroft, J. (1996) *Democracy and the Environment: Problems and Prospects* (Cheltenham: Edward Elgar).

Lang, T. and Gabriel, Y. (2005) 'A Brief History of Consumer Activism', in R. Harrison, T. Newholm and D. Shaw (eds) *The Ethical Consumer* (London: Sage).

Lash, S. (1994) 'Reflexivity and its Doubles. Structures, Aesthetics, Community', pp. 110–173 in U. Beck, A. Giddens and S. Lash *Reflexive Modernization. Politics, Tradition and Aesthetics in the Modern Social Order* (Cambridge: Polity Press).

Lathrop, K. V. (1991) 'Preempting Apples with Oranges: Federal Regulation of Organic Food Labelling', *J. Corp. L.*, 16, 885–930.

Laufer, W. S. (2003) 'Social Accountability and Corporate Greenwashing', *Journal of Business Ethics*, 43, 253–261.

Law, J. and Urry, J. (2004) 'Enacting the Social', *Economy and Society*, 33/3, 390–410.

Le Guillou, G. and Scharpé, A. (2000) *Organic Farming. Guide to Community Rules* (European Commission: Directorate-General for Agriculture).

Le Velly, R. (2007) 'Is large-scale Fair Trade possible?' pp. 201–215 in E. Zaccai (ed.) *Sustainable Consumption, Ecology and Fair Trade* (London: Routledge).

Levidow, L. and Boschert, K. (2008) 'Coexistence or contradiction? GM crops versus alternative agricultures in Europe', *Geoforum* 39, 174–190.

Lewis, A. and Mackenzie, C. (2000) 'Support for Investor Activism among U.K. Ethical Investors', *Journal of Business Ethics*, 24, 215–222.

Lidskog, R. and Sundqvist, G. (2004) 'From Consensus to Credibility. New Challenges for Policy-relevant Science', *Innovation*, 17, 205–226.

Liefferink, D., Andersen, M.S. and Enevoldsen, M. (2000) 'Interpreting Joint Environmental Policy-making: Between Deregulation and Political Modernization', in A. Mol, V. Lauber, and D. Liefferink (eds) *The Voluntary Approach to Environmental Policy: Joint Environmental Policy-making in Europe* (Oxford: Oxford University Press).

Lindahl, K. (2001) *Behind the Logo. The Development, Standards and Procedures of the Forest Stewardship Council and the Pan European Forest Certification Scheme in Sweden* (Gloucestershire: Fern).

Lindblom, C. (1977) *Politics and Markets. The World's Political-economic Systems* (New York: Basics books).

Lindblom, C. (2001) *The Market System* (New Haven: Yale University Press).

Lindén, A-L. and Carlson-Kanyama, A. (2007) 'Energy Efficiency in Residences – A Challenge for Women and Men', *Energy Policy*, 35, 2163–2172.

Lindén, A-L. and Klintman M. (2003) 'The Formation of Green Identities – Consumers and Providers', pp. 66–90 in A. Biel, B. Hansson and M. Mårtensson (eds), *Individual and Structural Determinants of Environmental Practice* (Aldershot: Ashgate).

Lindvert, J. (2006) *Ihålig Arbetsmarknadspolitik? Organisering och Legitimitet Igår och Idag* (Umeå: Borea).

Lindvert, J. (2008) 'The Political Logics of Accountability. From "Doing the Right Thing" to "Doing the Thing Right"' in M. Boström and C. Garsten (eds) *Organizing Transnational Accountability* (Cheltenham: Edward Elgar).

Lintott, J. (1998) 'Beyond the Economics of More: The Place of Consumption in Ecological Economics', *Ecological Economics*, 25, 239–248.

Locke, J. (1689/1997) *Two Treatises of Government*, M. Goldie (ed.) (London: Everyman).

Lovan, R., Murray, M. and Shaffer, R. (eds) (2004) *Participatory Governance: Planning, Conflict Mediation and Public Decision-Making in Civil Society* (Aldershot: Ashgate).

Lundqvist, L.J. (1996) 'Sweden', in P. M. Christiansen (ed.), *Governing the Environment: Politics, Policy, and Organization in the Nordic 'Countries'* (Copenhagen: Nordic Council of Ministers).

Lundström, T. and Wijkström, F. (1997) *The Nonprofit Sector In Sweden* (Manchester: Manchester University Press).

Lönn, M. (2003) *Konsumenternas Syn på Miljömärkt Vildfångad Fisk* (Uppsala: KRAV).

van Maanen, J. and Pentland, B. T. (1994) 'Cops and Auditors', in S. Sitkin and R. Bies (eds) *The Legalistic Organization* (London: Sage).

Macnaghten, P. and Urry, J. (1998) *Contested Natures* (London: Sage).

Magnusson, M. K., Arvola, A., Koivisto Hursti, U-K., Åberg, L. and Sjödén, P.-O. (2001) 'Attitudes Towards Organic Foods Among Swedish Consumers', *British Food Journal*, 103, 209–226.

Majone, G. (1996) 'Regulation and its modes', in G. Majone *Regulating Europe* (London: Routledge).

Marsden, T., Flynn, A. and Harrison, M. (2000) *Consuming Interests: The Social Provision of Foods* (London: UCL Press).

Martin, J. (1986) 'Happy Returns for Do-gooders', *Financial World*, 18 March, 32–33.

McAdam, D., McCarthy, J. D. and Zald, M. N. (eds) (1996) *Comparative Perspectives on Social Movements: Political Opportunities, Mobilizing Structures, and Cultural Framings* (Cambridge: Cambridge University Press).

McAvoy, S. (2000) 'Glickman Announces New Proposal for National Organic Standards', USDA Release No. 0074.00, http://www.usda.gov/news/releases/2000/03/00, accessed 18 December 2003.

McEachern, M. G. and Schroder, M. J. A. (2004) 'Integrating the Voice of the Consumer within the Value Chain: A Focus on Value-based Labelling Communications in the Fresh Meat Sector', *Journal of Consumer Marketing*, 21, 497–509.

McNichol, J. (2003) 'International NGO Certification Programs as New Para-Regulatory Forms? Lessons from a Frontrunner', Paper prepared for Conference on 'The Multiplicity of Regulatory Actors in the Transnational Space', Uppsala University, Department of Business Studies, 23–24 May 2003.

Meidinger, E. (1999) '"Private" environmental regulation, human rights and community', *Buffalo Environmental Law Journal*, 7, 123–237.

Mertig, A., Dunlap, R. and Morrison D. (2001) 'The Environmental Movement in the United States', in R. Dunlap and W. Michelson (eds) *Handbook of Environmental Sociology* (Westport CT: Greenwood Press).

Micheletti, M. (1995) *Civil Society and State Relations in Sweden* (Aldershot: Avebury).

Micheletti, M. (2003) *Political Virtue and Shopping: Individuals, Consumerism and Collective Action* (London/Basingstoke, UK: Palgrave Macmillan).

Micheletti, M. and Føllesdal, A. (2007) 'Shopping for Human Rights', *Journal of Consumer Policy*, 30, 167–175.

Micheletti, M. and Stolle, D. (2005) 'A Case of Discursive Political Consumerism: The Nike-Email Exchange', in M. Boström, A. Føllesdal, M. Klintman, M. Micheletti and M. P. Sørensen (eds) *Political Consumerism: Its Motivations, Power, and Conditions in the Nordic Countries and Elsewhere* (Copenhagen:

Nordic Council of Ministers). Electronically available at http://www.norden. org/pub/velfaerd/konsument/sk/TN2005517.pdf.

Micheletti, M., Follesdal, A. and Stolle, D. (eds) (2004) *Politics, Products, and Markets. Exploring Political Consumerism Past and Present* (New Brunswick, London: Transaction Publishers).

Michelsen, J. (2001) 'Recent Development and Political Acceptance of Organic Farming in Europe', *Sociologia Ruralis*, 41, 3–20.

Miller, H. I. (2007) 'Two Views of the Emperor's New Clones', *Nature Biotechnology*, 25(3), 281.

von Mises, L. E. (1944) *Omnipotent Government: The Rise of the Total State and Total War* (New Haven, CT: Yale University Press).

Mol, A., Lauber, V. and Liefferink, D. (eds) (2000) *The Voluntary Approach to Environmental Policy: Joint Environmental Policy-making in Europe* (Oxford: Oxford University Press).

Mont, O. and Bleischwitz, R. (2007) 'Sustainable Consumption and Resource Management in the Light of Life Cycle Thinking', *European Environment*, 17, 59–76.

Moore, O. (2006) 'Understanding Post-organic Fresh Fruit and Vegetable Consumers at Participatory Farmers' Markets in Ireland: Reflexivity, Trust and Social Movements', *International Journal of Consumer Studies*, 30, 416–426.

Moyer, W. and Josling, T. (2002) *Agricultural Policy Reform. Politics and Process in the EU and US in the 1990s* (Aldershot: Ashgate).

Mörth, U. (ed.) (2004) *Soft Law in Governance and Regulation: An Interdisciplinary Analysis* (Cheltenham, UK: Edward Elgar).

Nestle, M. (2002) *Food politics: how the food industry influences nutrition and health* (Berkeley: University of California Press).

Newell, P. (2005) 'Towards a Political Economy of Global Environmental Governance', pp. 187–201 in P. Dauvergne (ed.) *Handbook of Global Environmental Politics* (Northampton: Edward Elgar).

Nilsson, H., Tuncer, B. and Thidell, A. (2004) 'The Use of Eco-labelling Like Initiatives on Food Products to Promote Quality Assurance – Is there Enough Credibility?', *Journal of Cleaner Production*, 12, 517–527.

Nilsson, S. (2005) 'Egendeklaration Istället för Miljömärkning: En Reaktion mot Regelexplosionen', *Score working paper series 2005:5* (Stockholm: Score). Electronically available at http://www.score.su.se.

Nimon, W. and Beghin, J. (1999) 'Are Eco-labels Valuable? Evidence from the Apparel Industry', *American Journal of Agricultural Economics*, 81(4), 801–811.

Nordic Council of Ministers (1998) 'Vem Tar Ansvar för Fisket? – Rapport från Nordiskt Seminarium om Tvång och Legitimitet i Fiskeförvaltningen', TemaNord 1998: 506 (Copenhagen: Nordic Council of Ministers).

Nordic Council of Ministers (2000) *An Arrangement for the Voluntary Certification of Products of Sustainable Fishing* (Copenhagen: Nordic Council of Ministers).

Nordic Council of Ministers (2001) *Food Labelling: Nordic Consumers' Proposals for Improvements. A Pan-Nordic Survey of Consumer Behaviour and Attitudes towards Food Labelling.* TemaNord 2001: 573 (Copenhagen, Nordic Council of Ministers).

Nowotny, H., Scott, P. and Gibbons, M. (2001) *Re-thinking Science. Knowledge and the Public in an Age of Uncertainty* (Cambridge: Polity)

Olson, M. (1965/1971) *The Logic of Collective Action. Public Goods and the Theory of Groups* (London: Harvard University Press).

Oosterver, P. (2005) *Global Food Governance*. PhD dissertation (Wageningen: Wageningen University).

O'Rourke, A. (2005) 'The Message and Methods of Ethical Investment', *Journal of Cleaner Production*, 11, 683–693.

Organic Foods Production Act of 1990 (Section 2119).

Organic Trade Association (2002) 'Passion for Organic Products Sparks Nearly 6,000 Essays', http://www.ota.com/news/press/18.html, accessed 21 January 2004.

Organic Trade Association (2004) *Organic Manufacturer Market Survey* (Greenfield, Massachusetts: Organic Trade Association).

Orlikowski, W. J. (2007) 'Sociomaterial Practices: Exploring Technology at Work', *Organization Studies*, 28, 1435–1448.

Ott, S. (ed.) (2001) *The Nature of the Nonprofit Sector* (Boulder: Westview Press).

Ozinga, S. (2001) *Behind the Logo. An Environmental and Social Assessment of Forest Certification Schemes* (Gloucestershire: Fern).

Parviainen, J. and Frank, G. (2003) 'Protected Forests in Europe Approaches-Harmonising the Definitions for International Comparison and Forest Policy Making', *Journal of Environmental Management*, 67, 27–36.

Patterson, L. A. (2000) 'Biotechnology Policy. Regulating Risks and Risking Regulation', pp. 317–344 in H. Wallace and W. Wallace (eds) *Policy-Making in the EU*, 4th edition (Oxford: Oxford University Press).

Peattie, K. and Crane, A. (2005) 'Green Marketing: Legend, Myth, Farce or Prophesy', *Qualitative Market Research: An International Journal*, 8(4), 357–370.

Pellizzoni, L. (2004) 'Responsibility and Environmental Governance', *Environmental Politics*, 13, 541–565.

Pepper, D. (1996) *Modern Environmentalism: An Introduction* (London: Routledge).

Peuhkuri, T. and Jokinen, P. (1999) 'The Role of Knowledge and Spatial Contexts in Biodiversity Policies: A Sociological Perspective', *Biodiversity and Conservation*, 8, 133–147.

Pierre, J. and Peters, B. G. (2000) *Governance, Politics and the State* (Houndmills: Macmillan Press).

Porter, G., Brown, J. W. and Chasek, P. (2000) *Global Environmental Politics*, 3rd edition (Boulder, Westview Press).

Power, M. (1997) *The Audit Society. Rituals of Verification* (Oxford, Oxford University Press).

Power, M. (2000) 'The Audit Society – Second Thoughts', *International Journal of Auditing*, 4, 111–119.

Power, M. (2007) *Organized Uncertainty. Designing a World of Risk Management* (Oxford: Oxford University Press).

Radin, R. F. and Stevenson, W. B. (2006) 'Comparing Mutual Fund governance and Corporate Governance', *Corporate Governance: An International Review*, 14(5), 367–376.

Rametsteiner, E. (2002) 'The Role of Governments in Forest Certification: a Normative Analysis Based on New Institutional Economics Theories', *Forest Policy and Economics*, 4, 163–173.

Raynolds, L. (2000) 'Re-embedding Global Agriculture: The International Organic and Fair Trade Movements', *Agriculture and Human Values*, 17, 297–309.

Raynolds, L., Murray, D. and Heller, A. (2007) 'Regulating Sustainability in the Coffee Sector: A Comparative Analysis of Third-party Environmental and Social Certification Initiatives', *Agriculture and Human Values*, 24, 147–163.

Rein, M. and Schön, D. (1993) 'Reframing policy discourse', pp. 143–166 in F. Fischer and J. Forester (eds) *The Argumentative Turn in Policy Analysis and Planning* (Durham; London: Duke University Press).

Rémy, E. and Mougenot, C. (2002) 'Inventories and Maps: Cognitive Ways of Framing the Nature Policies in Europe', *Journal of Environmental Policy & Planning*, 4, 313–322.

Rhodes, R. A. W. (1997) *Understanding Governance. Policy Networks, Governance, Reflexivity and Accountability* (Maidenhead: Open University Press).

Rhodes, R. A. W. (2000) 'Governance and Public Administration', pp. 54–90 in J. Pierre (ed.) *Debating Governance. Authority, Steering, and Democracy* (Oxford: Oxford University Press).

Rhodes, S. P. and Brown, L. B. (1997) 'Consumers Look for the Ecolabel', *Forum for Applied Research and Public Policy*, 12, 109–115.

Ribbing, P. (2002), in Miljöeko 2002/5.

Richardson, J., Gustafsson, G. and Jordan, G. (1982) 'The Concept of Policy Style', in J. Richardson (ed.) *Policy Styles in Western Europe* (London: George Allen & Unwin).

Roff, R. J. (2007) 'Shopping for change? Neoliberalizing activism and the limits to eating non-GMO', *Agriculture and Human Values*, 24, 511–522.

Rosenau, J. N. (2003) *Distant Proximities. Dynamics Beyond Globalization* (Princeton: Princeton University Press).

Rosendal, K. (2005) 'Governing GMOs in the EU: A Deviant Case of Environmental Policy-making?', *Global Environmental Politics*, 5, 83–104.

van Rooy, A. (2004) *The Global Legitimacy Game. Civil society, Globalization, and Protest* (New York: Palgrave Macmillan).

Rubik, F. (2002) 'European Environmental Product Information Schemes (EPIS) and European Integrated Product Policy (IPP)' in F. Rubik and G. Scholl (eds) *Eco-labelling Practices in Europe. An Overview of Environmental Product Information Scheme*s. Schriftenreihe des IÖW 162/02, (Berlin: IÖW).

Rubik, F. and Scholl, G. (2002) 'Environmental Product Information Systems (EPIS) in the Member States of the European Union and Norway – Findings and Conclusions', in F. Rubik and G. Scholl (eds) *Eco-labelling Practices in Europe. An Overview of Environmental Product Information Schemes*, Schriftenreihe des IÖW 162/02 (Berlin: IÖW).

Rydén, R. (2005) *Marknaden, Miljön & Politiken: Småbrukarnas och Ekoböndernas Förutsättningar och Strategier 1967–2003* (Uppsala: Uppsala University, Dept of History).

Sabatier, P. and Jenkins-Smith, H. (1999) 'The Advocacy Coalition Framework: An Assessment', in P. Sabatier (ed.) *Theories of the Policy Process* (Boulder: Westview Press).

Sammer, K. and Wüstenhagen, R. (2006) 'The Influence of Eco-Labelling on Consumer Behaviour – Results of a Discrete Choice Analysis for Washing Machines', *Business Strategy and the Environment*, 15, 185–199.

Schmidt, S. and Werle, R. (1998) *Coordinating Technology. Studies in the International Standardization of Telecommunications* (Cambridge, Mass: The MIT Press).

Scholl, G. (2002) 'Environmental Product Information Schemes (EPIS) in Germany', in F. Rubik and G. Scholl (eds) *Eco-labelling Practices in Europe. An Overview of Environmental Product Information Schemes*, Schriftenreihe des IÖW 162/02 (Berlin: IÖW).

Schön, D. and Rein, M. (1994) *Frame Reflection* (New York: Basic Books).

Schumacher, E. F. (1973) *Small is Beautiful* (London: Blond and Briggs).

Schwartz, M. S. (2003) 'The Ethics of Ethical Investing', *Journal of Business Ethics*, 43(3), 195–213.

Seippel, Ø. (2007) 'Trees, Ecology and Biological Diversity: Norwegian Nature Protection and Environmentalism', in C. S. A. van Koppen and W. T. Markham (eds) *Protecting Nature: Organizations and Networks in Europe and the U.S.* (Cheltenham: Edward Elgar).

Silver, I. (1997) 'Constructing "Social Change" through Philanthropy: Boundary Framing and the Articulation of Vocabularies of Motives for Social Movement Participation', *Sociological Inquiry*, 67, 488–503.

Skillius, Å. (2005) *Etiska fonder 2005*, Research Paper http://folksam.se/resurser/pdf/Etikfondindex%202005.pdf, accessed 9 January 2008.

Sklar, K. (1998) 'The Consumers' White Label Campaign of the National Consumers' League, 1898–1918', pp. 17–36 in S. Strasser, C. McGovern and M. Judt (eds), *Getting and Spending: European and American Consumer Societies in the Twentieth Century* (Cambridge: Cambridge University Press).

Smith, A. (1776/1974) *The Wealth of Nations* (Harmondsworth: Penguin Books).

Snow, D. and Benford, R. (1988) 'Ideology, Frame Resonance, and Participant Mobilization', in B. Klandermans, H. Kriesi and S. Tarrow (eds) *From Structure to Action: Comparing Social Movement Research Across Cultures* (London: Jai Press).

Snow, D., Rochford, B. Jr., Worden, S. and Benford, R. (1986) 'Frame Alignment Processes, Micromobilization, and Movement Participation', *American Sociological Review*, 51, 464–481.

Social Investment Forum (2006) *2005 Report on Socially Responsible Investing Trends in the United States*, http://www.socialinvest.org/areas/research/trends/SRI_Trends_Report_2005.pdf, accessed 15 November 2006.

Soneryd, L. (2008) 'Regulating Coexistence: The Creation of New Discursive Sites for Battles over GM Crops', in R. Lidskog, L. Soneryd and Y. Uggla, *Making risk governable*, unpublished manuscript.

Soper, K. (1995) *What is Nature?* (Oxford: Blackwell).

Soper, K. and Trentmann, F. (2007) 'Introduction', in K. Soper and F. Trentmann (eds) *Citizenship and Consumption* (Houndmills: Palgrave Macmillan).

Sörbom, A. (2002) *Vart tar Politiken Vägen? Om Individualisering, Reflexivitet och Görbarhet i det Politiska Engagemanget* (Stockholm: Almqvist & Wiksell International).

Sørensen, M. P. (2004) *Den Politiske Forbruger – i det Liberale Samfund* (Copenhagen, Hans Reitzels Forlag).

SOU 1997/97, 'Skydd av Skogsmark. Behov och kostnader', Huvudbetänkande av Miljövårdsberedningen.

Spaargaren, G. and Mol, A. (1997) 'Sociology, Environment and Modernity: Ecological Modernization as a Theory of Social Change', in G. Spaargaren (ed.) *The Ecological Modernization of Production and Consumption. Essays in Environmental Sociology* (Wageningen: Wageningen University).

Sparkes, R. (2001) 'Ethical Investments: Whose Ethics, which Investment?', *Business Ethics: A European Review*, 10(3), 194–205.

Spencer, R. C. (2001) 'Assets in Socially Screened Investments Grew by 183%', *Employee Benefit Plan Review*, 56, 30–32.

Steinberg, M. (1998) 'Tilting the frame: Considerations on collective action framing from a discursive turn', *Theory and Society*, 27, 845–872.

Stø, E. (2002) 'Environmental Product Information Schemes (EPIS) in the Nordic Countries: Denmark, Sweden and Finland', in F. Rubik and G. Scholl (eds) *Eco-labelling Practices in Europe. An Overview of Environmental Product Information Schemes*, Schriftenreihe des IÖW 162/02 (Berlin: IÖW).

Stokke, O. S. (ed.) (2001) *Governing High Seas Fisheries. The Interplay of Global and Regional Regimes* (Oxford: Oxford University Press).

Stokke, O. S. and Coffey, C. (2004) 'Precaution, ICES and the Common Fisheries Policy: a Study of Regime Interplay', *Marine Policy*, 28, 117–126.

Stokke, O. S., Gulbrandsen, L. H., Hoel, A. H. and Braathen, J. (2005) 'Ecolabelling and Sustainable Management of Forestry and Fisheries: Does it Work?', in M. Boström, A. Føllesdal, M. Klintman, M. Micheletti and M. P. Sørensen (eds) *Political Consumerism: Its Motivations, Power, and Conditions in the Nordic Countries and Elsewhere* (Copenhagen: Nordic Council of Ministers). Electronically available at http://www.norden.org/pub/velfaerd/konsument/sk/TN2005517.pdf.

Strasser, S., McGovern, C. and Judt, M. (eds) (1998) *Getting and Spending: European and American Consumer Societies in the Twentieth Century* (Cambridge: Cambridge University Press).

Streeck, W. and Schmitter, P. (1985) 'Community, Market, State – and Associations? The Prospective Contribution of Interest Governance to Social Order', pp. 1–29 in W. Streeck and P. Schmitter (eds) *Private Interest Government. Beyond Market and State* (London: Sage).

Strømsnes, K. (2005) 'Political consumption in Norway: Who, why – and does it have any effect?' in M. Boström, A. Føllesdal, M. Klintman, M. Micheletti and M. P. Sørensen (eds) *Political Consumerism: Its Motivations, Power, and Conditions in the Nordic Countries and Elsewhere* (Copenhagen: Nordic Council of Ministers). Electronically available at http://www.norden.org/pub/velfaerd/konsument/sk/TN2005517.pdf.

Swedish National Board of Agriculture (2004) *Mål för Ekologisk Produktion 2010.* Rapport 2004:19.

Swesif (2007) *Welcome to Swesif/asset classes/statistics, trends and links.* http://www.swesif.org/eng/index.html, accessed 25 May 2007.

Swidler, A. (1986) 'Culture in Action: Symbols and Strategies', *American Sociological Review*, 51, 273–286.

Synnott, T. (2005) 'Some notes on the early years of FSC', unpublished manuscript (Bonn: FSC).

Tamm Hallström, K. (2004) *Organizing International Standardization: ISO and the IASC in Quest of Authority* (Cheltenham: Edward Elgar).

Tamm Hallström, K. (2008) 'ISO Expands its Business into Social Responsibility', in M. Boström and C. Garsten (eds) *Organizing Transnational Accountability* (Cheltenham: Edward Elgar).

van Tatenhove, J., Arts, B. and Leroy, P. (2000) *Political Modernization and the Environment: The Renewal of Environmental Policy Arrangement* (London: Kluwer Academic Publishers).

Teisl, M., Peavey, S., Newman, F., Buono, J. and Hermann, M. (2002) 'Consumer Reactions to Environmental Labels for Forest Products: A preliminary...', *Forest Products Journal*, 52, 44.

Terragni, L. and Kjærnes, U. (2005) 'Ethical Consumption in Norway: Why is it so low?', in M. Boström, A. Føllesdal, M. Klintman, M. Micheletti and M. P. Sørensen (eds) *Political Consumerism: Its Motivations, Power, and Conditions in the Nordic Countries and Elsewhere* (Copenhagen: Nordic Council of Ministers). Electronically available at http://www.norden.org/pub/velfaerd/konsument/sk/TN2005517.pdf.

The American Crop Protection Association (ACPA, 12 June 2000) 'Comments on National Organic Program Proposed Rule'. Electronically available at: http://www.croplifeamerica.org/public/issues/organic/natrule.html.

The Economist (2006) 'Voting with your trolley', 7 December 2006. Electronically available at http://www.economist.com/business/PrinterFriendly.cfm?story_id=8380592.

The Economist (May 19, 2007) 'Not on the Label: Why Adding "Carbon Footprint" Labels to Foods and Other Products is Tricky', 383 (8529), 90.

The European Union Eco-labelling Board (EUEB) (2004) 'Presidential Meeting of the European Union Eco-labelling Board', address by Mr Michael Ahern, T. D., Minister for Trade & Commerce, at the Presidential Meeting on Thursday 15 April 2004 in Dublin, ENDS/TC90.

Thedvall, R. (2006) *Eurocrats at Work. Negotiating Transparency in Postnational Employment Policy* (Stockholm: Almqvist & Wicksell International).

Thornber, K. (2003) 'Certification: A Discussion of Equity Issues', pp. 63–82 in E. Meidinger, C. Elliot and G. Oesten (eds) *Social and Political Dimensions of Forest Certification* (Remagen-Oberwinter: Forstbuch).

Tobiasen, M. (2005) 'Political Consumerism in Denmark', in M. Boström, A. Føllesdal, M. Klintman, M. Micheletti and M.P. Sørensen (eds) *Political Consumerism: Its Motivations, Power, and Conditions in the Nordic Countries and Elsewhere* (Copenhagen: Nordic Council of Ministers). Electronically available at http://www.norden.org/pub/velfaerd/konsument/sk/TN2005517.pdf.

Torjusen, H., Sangstad, S., O'Doherty Jensen, K. and Kjærnes, U. (2004) *European Consumers' Conceptions of Organic Food*, Professional report no. 4–2004 (Oslo: SIFO).

Törnqvist, T. (1995) *Skogsrikets Arvingar: En Sociologisk Studie av Skogsägarskapet inom Privat, Enskilt Skogsbruk* (Uppsala: SLU, Institutionen för Skog-Industri-Marknad Studier).

Trewavas, A. (2001) 'Urban Myths of Organic Farming', *Nature*, 410, 409–410.

Triandafyllidou, A. and Fotiou, A. (1998) 'Sustainability and Modernity in the European Union: A Frame Theory Approach to Policy-Making', *Sociological*

*Research Online*, 3(1), http://www.socresonline.org.uk/socresonline/3/1/2.
html, accessed 23 January 2003.

USDA Hearing on the Proposed Organic Rule, http://www.pmac.net/nospdf.
htm, accessed 19 December 2003.

USDA News Release (2000) *Glickman Announces National Standards For
Organic foods*, http://www.usda.gov/news/releases/2000/12/0425.htm,
accessed 23 January 2003.

US Department of Energy (2004) *Guide to Purchasing Green Power*, http://www.
epa.gov/greenpower/pdf/purchasing_guide_for_web.pdf, accessed 12 October
2007.

US Food and Drug Administration (1992) 'Statement of Policy: Foods Derived
from New Plant Varieties', *Federal Register*, 57(104), 22991.

Vachon, S. and Menz, F. C. (2006) 'The Role of Social, Political, and Economic
Interests in Promoting State Green Electricity Policies', *Environmental Science
and Policy*, 9(7–8), 652–662.

Vaupel, S. (1997) 'Advising Producers of Organic Crops', *Drake Journal of
Agricultural Law*, 2(1), 137–139.

Vogel, D. (2001) *The New Politics of Risk Regulation in Europe* (London: Centre for
Analysis of Risk and Regulation, LSE).

Vogel, D. (2004) 'Tracing the American Roots of the Political Consumerism
Movement', pp. 83–100 in M. Micheletti, A. Follesdal and D. Stolle (eds)
*Politics, Products, and Markets. Exploring Political Consumerism Past and Present*
(New Brunswick, London: Transaction Publishers).

Waide, P. (1998) *Monitoring of Energy Efficiency Trends of European Domestic
Refrigeration Appliances*, final report. PW Consulting for ADEME on behalf of
the European Commission (SAVE) (Manchester: PW Consulting).

Waide, P. (2001) *Monitoring of Energy Efficiency Trends of Refrigerators, Freezers,
Washing Machines and Washer-Driers Sold in the EU*, final report. PW
Consulting for ADEME on behalf of the European Commission (SAVE)
(Manchester: PW Consulting).

Wälti, S., Kübler, D. and Papadopoulos, Y. (2004) 'How Democratic is
"Governance"? Lessons from Swiss Drug Policy', *Governance: An International
Journal of Policy, Administrations, and Institutions*, 17, 83–113.

Warren, M. E. (1999) *Democracy & Trust* (Cambridge, UK: Cambridge University
Press).

Wasik, J. F. (1996) *Green Marketing & Management: A Global Perspective* (London:
Blackwell).

Weinberg, A. (1998) 'Distinguishing among Green Business: Growth, Green,
and Anomie', *Society and Natural Resources*, 11(3), 241–251.

Weir, A. (2000) 'Meeting Social and Environmental Objectives through
Partnership: the Experience of Unilever', pp. 118–124 in J. Bendell (ed.)
*Terms for Endearment. Business, NGOs and Sustainable Development* (Sheffield,
Greenleaf Publishing).

Worcester, R. and Dawkins, J. (2005) 'Surveying Ethical and Environmental
Attitudes', pp. 189–203 in R. Harrison, T. Newholm and D. Shaw (eds), *The
Ethical Consumer* (London: Sage).

Yearley, S. (1996) *Sociology, Environmentalism, Globalization* (London: Sage).

Yearley, S. (2005) *Making Sense of Science. Understanding the Social Study of Science*
(London: Sage).

Zaccai, E. (ed.) (2007) *Sustainable Consumption, Ecology and Fair Trade* (London: Routledge).

Zald, M. N. and McCarthy, J. D. (eds) (1997) *Social Movements in an Organizational Society: Collected Essays* (Oxford: Transaction Books).

Zavestoski, S., Shulman, S. and Schlosberg, D. (2006) *Science, Technology & Human Values* 31(4), 383–408.

Zinkhan, G. M. and Carlson, L. (1995) 'Green Advertising and The Reluctant Consumer', *Journal of Advertising*, 24(2), 1–6.

Zwick, D., Denegri-Knott, J. and Schroeder, J. E. (2007) 'The Social Pedagogy of Wall Street: Stock Trading as Political Activism?' *Journal of Consumer Policy*, 30, 177–199.

# Index

*Note*: Page numbers in **bold** denote tables and figures.

uncertainty
  and consumer insight, **40**, 41, 43,
    182, 194
  and framing, 114, 119, 128
  *see also* knowledge uncertainty;
    precaution; risk uncertainty
United Nations (UN), 32, 58, 63
United Nations Conference for
  Environment and Development
  (UNCED), 50–1, 122, 205
United Nations Environment
  Programme (UNEP), 50, 63
United States
  consumer activism, policy, and
    regulation, 17, 21–2
  and forest certification and
    labelling case, 50, 52–3
  and GM-labelling case, 53–5
  and green electricity labelling case,
    58–9
  and green mutual funds case, 64
  and organic food labelling case, 46,
    48–50
  and policy context, 14–15, 83–6,
    88–91, 93–4, 97–9, 101, 103,
    105, 110–11, 167–8, 183–6
Urry, J., 22, 207
US Department of Agriculture
  (USDA), 49–50, 73, 79, 91, 94, 98,
  118, 120, 122–3, 150, 160
US Department of Energy, 59–60
US Food and Drug Agency, 72, 94, 120
US Green Seal, 30, 206

Vachon, S. and Menz, F.C., 57, 58, 59
Vaupel, S. 105
Vogel, D., 21, 87, 93–4, 99, 167–8

Waide, P., 61
Wälti, S. and colleagues, 92, 210
Warren, M.E., 158
Wasik, J.F., 43
Weinberg, A., 23
Weir, A., 56
Whole Foods Market, 30, 147, 206
Worcester, R. and Dawkins, J., 36
World Conservation Union (IUCN),
  50
World Trade Organization
  (WTO), 72, 86, 95–7, 102, **145**,
  185, 206
World Wide Fund for Nature (WWF)
  and forest certification, 50, 51, 77,
    78, 81, 107, 139, 210
  and role of EMOs, 152, 154
  and seafood labelling, 56, 92, 107,
    141, 146, 158, 159, 208

Yearley, S., 4, 5, 43, 162

Zaccai, E., 10, 179, 180, 199
Zald, M.N., 103
Zavestoski, S. and colleagues, 8, 44
Zinkhan, G.M. and Carlson, L.,
  24, 37
Zutshi, A., 23, 32
Zwick, D. and colleagues, 34